D1238549

WITHDRAWN
UML LIBRARIES

OCCUPATIONAL HEALTH
AND
SAFETY

Terms, Definitions, and Abbreviations

SECOND EDITION

OCCUPATIONAL HEALTH AND SAFETY

Terms, Definitions, and Abbreviations

SECOND EDITION

Robert G. Confer
Thomas R. Confer

LEWIS PUBLISHERS

Boca Raton London New York Washington, D.C.

Library of Congress Cataloging-in-Publication Data

Confer, Robert G.
    Occupational health and safety : terms, definitions, and
abbreviations / Robert G. Confer, Thomas R, Confer. -- 2nd ed.
        p. cm.
    Includes bibliographical references.
    ISBN 1-56670-361-1 (alk. paper)
    1. Industrial hygiene--Dictionaries. 2. Industrial safety-
-Dictionaries. I. Confer, Thomas R. II. Title.
    RC963.A3C66 1999
    616.9′803′03—dc21   99-12086
                                                                     CIP

This book contains information obtained from authentic and highly regarded sources. Reprinted material is quoted with permission, and sources are indicated. A wide variety of references are listed. Reasonable efforts have been made to publish reliable data and information, but the author and the publisher cannot assume responsibility for the validity of all materials or for the consequences of their use.

Neither this book nor any part may be reproduced or transmitted in any form or by any means, electronic or mechanical, including photocopying, microfilming, and recording, or by any information storage or retrieval system, without prior permission in writing from the publisher.

The consent of CRC Press LLC does not extend to copying for general distribution, for promotion, for creating new works, or for resale. Specific permission must be obtained in writing from CRC Press LLC for such copying.

Direct all inquiries to CRC Press LLC, 2000 N.W. Corporate Blvd., Boca Raton, Florida 33431.

Trademark Notice: Product or corporate names may be trademarks or registered trademarks, and are only used for identification and explanation, without intent to infringe.

© 1999 by Robert G. Confer
Lewis Publishers is an imprint of CRC Press LLC

No claim to original U.S. Government works
International Standard Book Number 1-56670-361-1
Library of Congress Card Number 99-12086
Printed in the United States of America 1 2 3 4 5 6 7 8 9 0
Printed on acid-free paper

FOREWORD

The practice of industrial hygiene (IH) as well as the role of the industrial hygienist has been changing over the past several years. The scope of the industrial hygienist's daily activities has undergone the biggest change. Many of the employee exposure assessments and field work that an IH formerly carried out are now done by others. This is due, in many instances, to management's recognition that industrial hygienists should play a larger role in managing their functions, becoming involved with safety and environmental personnel in those activities, and becoming resources in process safety management, risk assessment, plan development for regulatory compliance, as well as in employee training program development, and the training of personnel on health and safety issues.

Downsizing in industry plus outsourcing have resulted in many industrial hygienists entering the consulting field, either in an entrepreneurial role or as employees of consulting firms. In both instances, they have had to broaden their knowledge in safety and environmental issues in order to provide the services that clients want. Providing only industrial hygiene support is not sufficient to attract and satisfy a client.

In the past industrial hygienists communicated primarily with physicians, nurses, toxicologists, analytical laboratory personnel, plant supervisors and engineers, and, to a lesser degree, management. Today they are working closely with safety, environmental, training, and human resources personnel, and are communicating on an almost daily basis with management, lawyers, process/engineering, and regulatory affairs, while still retaining a relationship and communication with their former contacts. Communications has become a most important factor in the daily routine of the industrial hygienist.

We have expanded this work in recognition of the changes in the IH field. It is our hope that it will serve not only the industrial hygiene community, but also, those of safety, environmental, regulatory affairs, human resources, engineering, and other disciplines.

AUTHORS

Robert Confer has worked in the disciplines of industrial hygiene, environmental health, and radiological health for over 40 years. His employers have included the Pennsylvania Department of Health, Westinghouse Electric Corporation, and Exxon Corporation. Since retiring from Exxon in 1989, he has worked as a private consultant to a number of U.S. based companies.

Through work assignments in more than 35 states and 30 foreign countries, Confer has addressed occupational, radiological, and environmental health concerns, as well as the training of personnel in the practice of field industrial hygiene.

Robert Confer has contributed chapters for several books, authored 27 articles that have been published in peer-reviewed journals, written two books, and received three patents.

Confer no longer climbs towers, enters tanks, rides ships or barges, performs stack sampling, or works turn-arounds. His present activities, for the most part, include conducting indoor air quality studies, training employees with respect to hazard communications, preparing written compliance programs, conducting industrial hygiene program audits and qualitative exposure assessments, reviewing and commenting on depositions, and giving depositions.

Thomas R. Confer has been employed in the environmental and industrial hygiene fields for 24 years. He began his career as an environmental chemist in South Carolina. In that capacity he conducted field studies for airborne and soluble environmental contaminants. His initial industrial hygiene experience was with Exxon Company's Bayway Refinery.

Confer spent two years with the U.S. Occupational Safety and Health Administration in South Carolina as a field industrial hygienist. He was employed for five years by International Flavors and Fragrances as corporate environmental health and safety manager, and was employed by BASF Corporation as manager of environmental health and safety in California. He is currently president of Integrated Environmental Systems, a national environmental engineering and consulting company.

Confer's work has included comprehensive industrial hygiene surveys, the development and management of underground storage tank removal programs, soil/groundwater contamination remediation, indoor air quality investigations and remediation, and development/management of assorted industrial hygiene sampling strategy programs. He has authored five publications and is active in various environmental and management organizations in California as well as the rest of the U.S.

A

a   atto, 1 E $^{-18}$.

A   ampere(s).

A1 Carcinogen   Confirmed Human Carcinogen (ACGIH)   A confirmed human carcinogen as classified by the ACGIH TLV Committee. Includes agents that are carcinogenic to humans based on the weight of evidence from epidemiologic studies of, or convincing clinical evidence in, exposed humans.

A2 Carcinogen   Suspected Human Carcinogen (ACGIH)   A suspected human carcinogen as classified by the ACGIH TLV Committee. Includes agents carcinogenic in experimental animals at dose levels, by route(s) of administration, at site(s), of histologic type(s), or by mechanism(s) that are considered relevant to worker exposure. Available epidemiologic studies are conflicting or insufficient to confirm an increased risk of cancer in exposed humans.

A3 Carcinogen   Animal Carcinogen (ACGIH)   An animal carcinogen as classified by the ACGIH TLV Committee. Includes agents that are carcinogenic in experimental animals at a relatively high dose, by route(s) of administration, at site(s), of histologic type(s), or by mechanism(s) that are not considered relevant to worker exposure. Available epidemiologic studies do not confirm an increased risk of cancer in exposed humans. Available evidence suggests that the agent is not likely to cause cancer in humans except under uncommon or unlikely routes or levels of exposure.

A4 Carcinogen   Not Classified as a Human Carcinogen (ACGIH)   Not classified as a human carcinogen by the ACGIH TLV Committee. There are inadequate data on which to classify these agents in terms of their carcinogenicity in humans and/or animals.

A5 Carcinogen   Not Suspected as a Human Carcinogen (ACGIH)   Not classified or suspected as a human carcinogen by the ACGIH TLV Committee. Includes agents that are not suspected to be human carcinogens on the basis of properly conducted epidemiologic studies in humans. These studies have sufficiently long follow-ups, reliable exposure histories, sufficiently high doses, and adequate statistical power to conclude that exposure to the agent does not convey a significant risk of cancer to humans. Evidence suggesting a lack of carcinogenicity in experimental animals will be considered if it is supported by other relevant data.

AA   Atomic absorption.

AAEE   American Academy of Environmental Engineers.

AAIH   American Academy of Industrial Hygiene.

AAL   Ambient air level.

AAOHN   American Association of Occupational Health Nurses.

AAOM   American Association of Occupational Medicine.

AAOO   American Academy of Ophthalmology and Otolaryngology.

AAP   Asbestos action plan.

AAQS   Ambient air quality standard.

AAR   Asbestos Analysts Registry.

AAS   Atomic absorption spectroscopy.

abate   To reduce in amount, degree, or intensity.

abatement   The act of abating or reducing a problem.

abatement (Air Pollution)   The reduction in the intensity or concentration of an ambient air pollutant.

abatement (Asbestos)   Control of the release of fibers from a source of asbestos-containing material during removal, enclosure, or encapsulation.

aberrant   Something that deviates substantially from that which is considered typical or normal.

aberration   A deviation from the norm.

ABIH   American Board of Industrial Hygiene.

abiotic   The nonliving components of ecosystems. Pertaining to or characterized by absence of life and being antagonistic to life.

ABOHN   American Board of Occupational Health Nurses Inc.

abrasive   A collection of discrete, solid particles that, when impinged on a surface, cleans, removes surface coatings, improves the quality of, or otherwise prepares to modify the characteristics of that surface, either by impact or friction.

abrasive blasting   *See* abrasive cleaning.

abrasive-blasting respirator   A continuous flow air-line respirator constructed to cover the wearer's head, neck, and shoulders to protect the wearer from rebounding abrasive.

abrasive cleaning   Process of cleaning surfaces by using materials such as sand, alumina, steel shot, walnut shells, etc., in a stream of high pressure air or water.

abs   Absolute.

abscissa   The distance along the horizontal coordinate, or x-axis, of a point from the vertical, or y-axis, of a graph.

absolute humidity   The weight of water vapor per unit volume of air (e.g., pounds per cubic foot or milligrams per cubic meter).

absolute pressure   Pressure measured with respect to zero pressure or a vacuum. It is equal to the sum of a pressure gauge reading and the atmospheric pressure at the measurement location.

absolute risk   The observed or calculated probability of an event occurring in a population under study. An estimate of health risk reached by determining the risk of occurrence of the disease in a study population.

absolute scale   A temperature scale based on absolute zero.

absolute temperature   Temperature based on an absolute scale expressed in either degrees Kelvin (K) or degrees Rankine (R), corresponding respectively to the centigrade (C) or (F) scales. Degrees Kelvin is obtained by adding 273 to the centigrade temperature, or subtracting the centigrade temperature, from 273 if below 0°C. Degrees Rankine are obtained by algebraically adding the Fahrenheit reading to 460. Zero degrees K is equal to -273°C, and 0 degrees R is equal to -459.69°F.

absolute zero   The minimum point on the thermodynamic temperature scale, which is 0 degrees Kelvin, -273.16 degrees centigrade, 0 degrees Rankine, or $-459.69°$ Fahrenheit. This is a hypothetical temperature at which there is a total absence of heat.

absorb   The penetration of one substance into the body of another.

absorbance   Logarithm to the base 10 of the transmittance.

absorbed dose (Industrial Hygiene)   The amount of a substance entering the body by crossing an exchange barrier (e.g., lungs, skin, etc.) as a result of an exposure to the material.

absorbed dose (Radiation)   The amount of energy imparted to matter by ionizing radiation per unit mass of irradiated material at the point of interest.

absorbent   A substance that takes in and absorbs other materials.

absorption (Acoustics)   The conversion of acoustical energy to heat or another form of energy within the medium of the sound-absorbing material.

absorption (Chemical)   The penetration of one substance into the inner structure of another. Physicochemical absorption occurs between a liquid and a gas or vapor by the assimilation of one material into another.

absorption (Physiologic)   The transfer of a substance across an exchange barrier of an organism (e.g., lungs, skin, etc.) and ultimately into body fluid and tissue.

absorption (Radiation)   The process by which the number of particles or photons entering a body of matter is reduced or attenuated by interaction with the matter.

absorption coefficient (Acoustics)   The fraction of incident sound absorbed or otherwise not reflected by a surface.

absorptive muffler (Acoustics)   A type of acoustic muffler that is designed to absorb sound energy as sound waves pass through it.

ac   Air changes.

ac   Alternating current.

AC   Air conditioner.

acanthamoeba   Organism that can develop in potable water as well as in the water of portable eye wash stations/units.

acaricide   A pesticide for destroying mites on animals, crops, and humans.

acarid   Any arachnid, such as ticks and mites.

ACBM   Asbestos-containing building material.

acceleration (Vibration)   The time rate of change of velocity, i.e., the rate at which the velocity of the vibratory motion changes direction.

acceleration loss (Ventilation)   The energy required to accelerate air to a higher velocity.

accelerator   A device for imparting kinetic energy to electrically charged particles such as electrons, protons, helium ions, and other ions of elements of interest. Common types of accelerators include the Van der Graaf, Cockcroft-Walton, cyclotrons, betatrons, and linear accelerators.

accelerometer   An instrument for measuring vibration.

acceptability (Instrument)   The willingness of personnel to use an instrument after considering its characteristics, such as weight, noise, response time, drift, portability, reliability, interference effects, etc.

acceptable entry conditions (Confined Space Entry)   Conditions that must exist in a confined space before entry is permitted to ensure that personnel can safely enter and carry out the work.

acceptable indoor air quality   Indoor air in which there are no known contaminants at harmful levels and with whose quality 80 percent of the occupants of the indoor environment are satisfied.

acceptable lift   [AL]   Ninety pounds multiplied by a series of factors related to the location of the object to be lifted, its distance from a specific position, and the lift frequency.

acceptable risk   A level of risk that is deemed by society to be acceptable and at which a seriously adverse result is highly unlikely to occur.

accessible (Asbestos-Containing Material)   A material that is subject to disturbance by occupants, custodians, and maintenance personnel in the course of their normal activities.

accident   An unplanned and uncontrolled event that may be injurious or damaging to an individual, property, or an operation. Any unplanned event that interrupts or interferes with the orderly progress of a production activity or process.

accidental release (Clean Air Act)   An unanticipated emission into the ambient air from a stationary source.

accident prevention   Efforts or countermeasures that are taken to reduce the number and severity of accidents.

accident rate   The accident experience relative to a base unit of measure (e.g., the number of disabling injuries per 100,000 person-hours worked).

accident site   The location of an unexpected occurrence, failure of equipment, or release of a hazardous material.

acclimation   *See* acclimatization.

acclimatization   An adaptive process which results in a reduction of the physiological response produced as a result of the application of a constant environmental stress, such as heat, on the body. The process of becoming accustomed to new conditions.

accommodation   The ability of the eye to focus for varying distances.

accountable   To be required to account for accomplishments or non-accomplishments relative to an assigned function or task. To be responsible to carry out an assignment given by management.

accreditation   A process by which an organization evaluates a program, institution, or an individual to determine if it meets predetermined standards.

accredited laboratory   Certification awarded to an analytical laboratory that has successfully participated in a proficiency testing program, such as that of the American Industrial Hygiene Association.

accuracy   The degree of agreement between a measured value and the accepted reference value, or the agreement of an instrument reading or analytical result to the true value. When referring to an instrument, accuracy represents the ability of the device to indicate the true value of the measured quantity. This is often expressed as a percentage of the full-scale range of the instrument.

ACD   Allergic contact dermatitis.

ACE (Indoor Air Quality)   Air change effectiveness.

acetylcholinesterase enzyme   [AChE]   An enzyme that is present in nervous tissue, muscle, and red cells that catalyzes the hydrolysis of acetylcholine to choline and acetic acid.

acfm   Actual cubic ft per minute.

ACGIH   American Conference of Governmental Industrial Hygienists.

AChE enzyme   Acetylcholinesterase enzyme.

ac/hr   Air changes per hour.

acid   Any chemical compound, one element of which is hydrogen, that dissociates in solution to produce free hydrogen ions.

acid gas   A gas that forms an acid when mixed with water.

acidosis   A pathologic condition resulting from the accumulation of acid in, or loss of base from, the body.

acid rain   The acidity in rain or snow that results from the oxidation of carbon, sulfur, or nitrogen compounds in the air and their subsequent absorption into the precipitation, thereby making it acidic.

ACM   Asbestos-containing material.

ac/min   Air changes per minute.

acne   An inflammatory disease of the sebaceous glands of the skin.

acoustic   Having properties, dimensions, or physical characteristics associated with sound waves.

acoustical insulation   Material designed to absorb noise energy that is incident upon it.

acoustical treatment   The use of acoustical (sound) absorbents, acoustical isolation, or other changes or additions to a noise source to improve the acoustical environment.

acoustic panel   Panel made of sound-absorbing material that reduces the reinforcement of sound by stopping the reflection of sound waves.

acoustics   The study of sound, including its generation, transmission, and effects.

acoustic trauma   Hearing loss as a result of sudden loud noise or blow to the head.

acquired immunity   Immunity that is not inherited.

acquired immunodeficiency syndrome   [AIDS]   The late clinical stage of infection with human immunodeficiency virus (HIV), recognized as a distinct syndrome.

acrid material   A substance that is bitter or sharp tasting or is stinging or irritating to the sense of taste and/or smell.

acro   Referencing to an extremity of the body, such as the end of bone tissue.

acro-osteolysis   A loss of calcium in the bones of the fingers and toes. Observed among workers exposed to vinyl chloride monomer, particularly among those cleaning PVC reactors by hand.

ACS   American Chemical Society.

ACS   Asbestos contaminated soil.

actinic   Pertaining to light rays beyond the violet end of the visible spectrum and that produce chemical effects.

actinic keratoses   Premalignant, erythematous, rough, scaly plaques that may develop into squamous cell carcinoma if left untreated.

actinic skin   Dry, brown, inelastic skin resulting from exposure to sunlight. Also referred to as sailor's skin or farmer's skin.

actinium series   Isotopes which belong to a chain of successive radioactive decays beginning with uranium 235 and ending with lead 207. All the decay products are metals, and those with atomic numbers higher than 92 are called transuranic elements. This series is also referred to as the actinide series.

action level   [AL]   The concentration of a substance to which a worker is exposed and at which specific actions or control measures are to be implemented. It is a term used by OSHA and NIOSH to express the level of exposure that triggers medical surveillance and other administrative controls. The action level is typically set at one-half the permissible exposure limit.

activated charcoal   A form of carbon characterized by a high absorption and adsorptive capacity for gases and vapors. Charcoal is activated by heating it with steam to 800–900°C. Commonly used in gas canisters and cartridges as a gas-adsorbent material and as an organic vapor/gas sampling media in industrial hygiene.

activated sludge   Sludge that has been aerated and subjected to bacterial action.

activation (Ionizing Radiation)   The process of making a material radioactive by bombardment with neutrons, protons, or other nuclear radiation; or simply the process of inducing radioactivity by irradiation.

active immunization   *See* immunization, active.

active sampling   An air sampling method in which air is drawn into the sampler where it is exposed to a sensor that measures the concentration of the contaminant in the sampled air, or is absorbed/adsorbed by a sorbent for later analysis. Also referred to as a pumped sample.

activity (Ionizing Radiation)   The number of nuclear transformations occurring in a given quantity of material per unit of time. The units of activity are the curie (Ci) and the becquerel (Bq).

act of God   An extraordinary, unpredicted, and unpreventable result of a normal activity.

acuity   Acuteness or sharpness of the senses, such as visual or hearing acuity. The sensitivity of receptors used in hearing or vision.

acute   Having a sudden onset and reaching a crisis rapidly. Effects are observed in a short period of time following exposure to an acute toxicant.

acute condition   Any condition characterized by rapid onset, short duration, and pronounced or severe symptoms.

acute dermal LD 50   The single dermal dose of a substance, expressed as milligrams per kilogram of body weight, that is lethal to 50 percent of the test population of animals under specified test conditions.

acute dose   A quantity which is administered in one day or as a single dose.

acute effect   A health effect that results following a brief exposure to a chemical, biological, or physical agent. For example, contact with phenol can cause death within a short period of time following skin contact with the liquid if a sufficient area is contaminated.

acute exposure   Exposure to a substance over a short period of time (e.g., minutes to hours).

acute LC 50   A concentration of a substance, expressed as parts per million parts of medium, that is lethal to 50 percent of the test population of animals under specified test conditions.

acute myelogenous leukemia   [AML]   A form of leukemia that can be a result of benzene exposure.

acute oral LD 50   A single orally administered dose of a substance, expressed as milligrams per kilogram of body weight, that is lethal to 50 percent of the test population of animals under specified test conditions.

acute radiation syndrome   A medical term for radiation sickness.

acute reaction   A sudden physiologic response as a result of an exposure to a hazard (i.e., chemical, physical, biological, ergonomic, etc.).

acute toxicity   An adverse health effect following an acute exposure to a toxic substance. The effect caused by a single dose or a short exposure to a toxic substance.

acute trauma   Injuries that occur suddenly and are immediately apparent, such as a laceration, burn, electrocution, etc., that can be linked to a well-defined accident.

acylation   The introduction of an acid radical into the molecule of a chemical compound.

ADA   Americans with Disabilities Act.

adaptation   A change in the structure or habits of an organism which enables it to better adapt to its surroundings. The process of accommodation to change.

additive   A chemical which is added in small quantities to improve the performance of some products, such as gasoline.

additive effect   A response in which the combined effect of two chemicals together is greater than the sum of the agents acting alone.

adduct   An unbonded association of two molecules in which a molecule of one component is either wholly or partly locked within the crystal lattice of the other.

adenocarcinoma   Carcinoma derived from glandular tissue or in which the tumor cells form recognizable glandular structures.

adenoma   A benign epithelial tumor in which the cells form recognizable glandular structures or in which the cells are clearly derived from glandular epithelium.

adhesion   The force or forces that cause two materials to stick together.

adiabatic   A change occurring in a body without gain or loss of heat from the surroundings.

adiabatic change   A change in the volume or pressure of a gas that occurs without a gain or loss of heat.

adiabatic heating   The heating of a gas as a result of compressing it.

adiabatic lapse rate   The change in temperature (i.e., $-5.4°$ F) for each 1000-ft increase in altitude.

adjudicate   To carry a lawsuit to a conclusion.

adjuvant   An additive which increases the effectiveness of the active ingredient.

administrative controls   Methods of controlling employee exposure to contaminants or physical stresses by rotating jobs, varying work assignments, increasing time periods away from the contaminant, changing a procedure or the time of day that a task is performed, etc.

administrative hearing   A hearing held by an authorized agency of the executive branch of the government to correct a violation of the rules or regulations of that agency.

admonition   A reprimand from a judge.

adsorb   The attraction and retention of a gas, liquid, or vapor on the surface of a solid or liquid.

adsorbate   The solid, liquid, or gas that is adsorbed.

adsorbent   A solid or liquid that absorbs other materials.

adsorption   The taking up, adherence, or condensation of a material, such as a gas or vapor, to the surface of another substance called the adsorbent, which is typically a solid. The condensation of gases, liquids, or dissolved substances on the surface of solids.

adulterate   To make impure, spurious, or inferior by adding extraneous or improper ingredients.

Advanced Notice of Proposed Rule-Making   [ANPRM]   A notice appearing in the Federal Register indicating the intention of a government agency to develop a regulation on the issue indicated in the notification.

adverse level (Air Pollution)   The concentration that results in the first discernible symptoms of discomfort relating to air pollution in humans.

adverse reaction   An undesirable or unwanted consequence of a preventive, diagnostic, or therapeutic procedure.

AEC   [U.S. Atomic Energy Commission]   The former name of the Nuclear Regulatory Commission.

aerate   To bring in contact with air.

aeration   To circulate oxygen (i.e., in air) through a substance.

aerobe   Microorganism that requires air for growth.

aerobic   Requiring free atmospheric oxygen for normal activity.

aerobic bacteria   Bacteria that require free oxygen for their life processes.

aerobic capacity   A measure of the ability of the pulmonary system to deliver sufficient oxygen to the blood and to remove carbon dioxide from it.

aerodynamic diameter   The diameter of a unit density sphere having the same settling velocity as the particle in question of whatever shape and density. It is also referred to as equivalent diameter.

aerometric   Pertaining to the measurement of the properties or contaminants of air.

aerosol   A fine suspension in air of liquid or solid particles that are sufficiently small in size to remain airborne and, frequently, to be respirable. An assemblage of small solid or liquid particles suspended in air.

aerosol photometer   An instrument used for detecting aerosols (i.e., dusts, mists, fumes, etc.) by exposing them to a source of illumination, typically a beam of light, as they are drawn through an enclosed volume, and measuring the scattered light created by the aerosol as it passes through the light beam.

affidavit   A written declaration made under oath before an authorized person, such as a notary public.

affinity   A chemical or physical attraction or attractive force.

affliction   Ailment, infirmity, or disease.

AGA   American Gas Association.

agar   A gel for the preparation of a solid culture media for the study of microorganisms.

age adjusted rate   The number of events occurring to a given age group, divided by the population at risk of that age group, times 100.

agent of disease   A factor, such as a microorganism, chemical substance, or form of radiation, whose presence or absence is essential for the occurrence of a disease.

age specific death rate (Epidemiology)   The mortality rate for a specified age group of a population.

age specific rate (Epidemiology)   A rate for a specific age group.

age standardization   A procedure for adjusting rates in order to minimize the effects of differences in age when comparing rates for different populations.

agglomeration   The consolidation of finer particles into larger ones by means of agitation, Brownian motion, or other forces. It is usually achieved by the neutralization of electric charges.

aggressive sampling   A sampling procedure employed following asbestos removal activities to demonstrate that the area is not contaminated with materials that contain asbestos fibers. It typically involves stirring up the air in the abated area to produce worst-case conditions, collecting air samples during this procedure, and analyzing the samples to determine the airborne level of asbestos fibers as structures per cubic centimeter.

aging (Environmental)   The gradual deterioration of a material due to long exposure to the environment.

Agreement State (Ionizing Radiation)   A state that has signed an agreement with the U.S. Nuclear Regulatory Commission allowing the state to regulate certain activities for the use of radioactive materials not normally regulated by the state, such as byproduct, source, and small quantities of nuclear material.

A-h   Ampere-hour.

AHERA   Asbestos Hazard and Emergency Response Act.

AHP   Air horsepower.

AHS   Air handling system.

AHU   Air handling unit.

AIA   Asbestos Information Association.

AIChE   American Institute of Chemical Engineers.

AIDS   *See* acquired immunodeficiency syndrome.

AIHA   American Industrial Hygiene Association.

AIHA Accredited Laboratory   A certification given by the AIHA to an analytical laboratory that has met specific requirements and successfully participated in the Proficiency Analytical Testing Program for quality control as established by the National Institute for Occupational Safety and Health.

AIHALAP   American Industrial Hygiene Association Laboratory Accreditation Program.

AIHC   American Industrial Health Council.

AIHC   American Industrial Hygiene Conference.

ailment   A condition of mild illness, physical disorder, or chronic disease.

AIPE   American Institute of Plant Engineers.

air   The mixture of gases that surround the earth. The major constituents of air are nitrogen (78.08 percent), oxygen (20.95 percent), argon (0.93 percent), and carbon dioxide (0.03 percent).

air-bone gap   The decibel difference in the hearing ability level at a particular frequency as determined by air conduction and bone conduction audiometric testing.

airborne asbestos sample   A sample that has been collected in a prescribed manner for determining the concentration of asbestos fibers or structures in the air by a specific analytic method (e.g., PCM or TEM).

airborne contamination limits (Radioactivity)   Concentration limits in picocuries per cubic centimeter of air that have been established to prevent radioactive overexposure of any organ of the body as a result of breathing air contaminated with radionuclides.

airborne dust   Airborne particulates, including the total dust and the respirable dust, present in the air.

airborne infection   A mechanism of transmission of an infectious agent by particles, dust, or droplet nuclei.

airborne microorganisms   Biologically active microorganisms that are suspended in air as fine particles or are attached to the surface of other suspended particulates.

airborne pathogen   A disease-causing microorganism that is transported through the ambient air or on particles present in the air.

airborne radioactive material   Radioactive material dispersed in the air in the form of a dust, fume, mist, vapor, gas, or other form.

airborne radioactivity area   An area where the concentration of airborne radioactivity is 10 percent above natural background.

airborne sound   Sound transmitted through the medium of air or gas rather than through a liquid or solid.

air change effectiveness (Indoor Air Quality)   [ACE]   The percentage of outside air that makes it from the air diffuser to the occupant level.

air changes per hour   [ac/hr]   The number of complete volumetric air changes in a space in a particular hour.

air changes per minute   [ac/min]   The number of complete volumetric air changes in a space in a particular minute.

air cleaner   A device designed to remove airborne contaminants, such as dusts, fumes, vapors, gases, etc., from the air.

air cleaning system   A device or a combination of equipment that is operated to reduce the concentration of airborne contaminants in the air supplied to an area/building or to remove contaminants from air being discharged into the ambient air.

air conditioning   A process of treating air to control factors such as temperature, humidity, and cleanliness, and to distribute the air throughout a space to meet the requirements of personnel and equipment.

air conduction (Acoustics)   The process by which sound is conducted through the air to the inner ear, with the outer ear canal serving as part of the pathway.

air contamination   Introduction of a foreign substance into the air, thus making the air impure.

air curtain (Oil Spill Control)   Bubbling air through a perforated pipe to cause an upward flow of water which slows the spread of oil.

air curtain (Ventilation)   Directing a flow of air across an opening, such as a doorway, to reduce air exchange through the opening or the entry of flying insects.

air erosion (Asbestos-Containing Material)   The movement of air over asbestos-containing building materials which may result in the release of asbestos fibers.

air exchange rate   The number of times that the outdoor air replaces the volume of air in a building per unit time, typically expressed as air changes per hour; or the number of times that the ventilation system replaces the air in a room or area in a building within a stated period.

air filter   A device for removing particulate matter from air. Designed to remove low dust concentrations from air passed through a filter.

airfoil sill/jamb   A tapered opening on the bottom (sill) and sides (jamb) of laboratory type hoods.

air gap   The vertical distance between the lowest opening from any pipe or faucet supplying potable water to a tank, sink, or other device and the flood level rim of the receptacle. The air gap should be two times the diameter of the water delivery pipe.

air horsepower   The theoretical horsepower required to drive a fan if there were no losses in the fan (i.e., if its efficiency were 100 percent).

air infiltration   The uncontrolled leakage of air into a building through cracks, open windows, holes, etc., from the building being under negative pressure and/or the influence of wind or temperature differences.

airline respirator   A respiratory protective device that is supplied for breathing air through a hoseline.

air lock   A system of enclosures or doors enabling ingress and egress and preventing the transfer of air between one area and an adjacent one.

air mass   A body of air covering hundreds of square miles in which the conditions of temperature, pressure, and moisture are similar throughout in a horizontal direction.

air monitoring   *See* air sampling.

Air Movement and Control Association   [AMCA]   An association which establishes performance classes for various types of fans.

air mover   Any type of device that is used to transfer air from one space/area to another.

air plenum   An air compartment connected to one or more ducts or to a slot-type hood for equalizing air distribution. Also referred to as a pressure-equalizing chamber.

air pollutant   Dust, fume, mist, smoke, or other aerosol, vapor, gas, or odorous substance, or any combination of these, which is emitted into the air or otherwise enters the ambient air.

air pollution   The presence of unwanted material in the air, such as dusts, vapors, smoke, etc., in sufficient concentration to affect the comfort, health, or welfare of residents or damage property exposed to the contaminated air. The deterioration of the quality of the air that results from the addition of impurities.

air pollution episode   A period of high concentration of air contamination, generally due to a temperature inversion and low wind conditions in an area.

air-purifying respirator   Respirator that removes a contaminant from the air being inhaled by the wearer by passing air through a filter or cartridge containing a solid sorbent, such as activated charcoal, before the air is inhaled.

air quality assessment   The collection, handling, analysis, and evaluation of air monitoring data to determine the quality of the air relative to accepted standards and to identify likely sources and the potential effects on people and the environment.

air quality control region   [AQCR]   A specific geographic area subject to air quality control measures. The EPA has divided the U.S. into a number of air quality control regions (e.g., 237). If a region does not meet emission requirements, it is classified as a non-attainment region.

air quality criteria   The amount of contaminant in air and the length of exposure to it that will result in adverse effects to health, welfare, or to the comfort of persons breathing it.

air quality index   A standardized indication of pollution intensity formulated by the EPA for public use. It is based on a combination of the environmental levels of ozone, suspended particulates, sulfur dioxide, and carbon monoxide. Also referred to as the air pollution index.

air quality standard   The regulated concentration of a pollutant that cannot be exceeded during a specified time period within a defined area. The concentrations of air pollutants that are considered tolerable.

air-reactive material   A material that will ignite at normal temperature when exposed to air.

air sampling   The collection of samples of air to determine the presence of and the concentration of a contaminant, such as a chemical, aerosol, or radioactive material, airborne microorganism, or other substance. The sample is analyzed to determine the amount present, and the concentration is calculated based on the sample volume.

airshed   A geographical area that shares the same air mass due to topography, meteorology, and climate.

air, standard (Ventilation)   Dry air at 70°F and 29.92 inches of mercury barometric pressure. It is equivalent to 0.075 pounds per cubic foot.

air, standard (Industrial Hygiene)   Dry air at 75°F and 29.92 inches of mercury barometric pressure.

air-supplied respirator   A respiratory protective device that provides a supply of breathable air from a source outside the contaminated work area. Includes airline respirators and self-contained respirators.

air-supply device   A hand- or motor-operated blower for supplying air to a hose–mask type respirator, or a compressor or other source of respirable air (e.g., breathing air cylinder) used for supplying air to the airline respirator.

air toxins   Chemical compounds that have been established as hazardous to human health. Also referred to as hazardous volatile organic compounds, including hydrocarbons such as benzene, halohydrocarbons such as carbon tetrachloride, nitrogen compounds such as amines, oxygen compounds such as ethylene oxide, and others.

air volume (Sampling)   The total volume of air passed through a sampling medium as determined by multiplying the flow rate in mL/min or L/min by the sampling time in minutes.

airway   The passage through which air enters and leaves the lungs.

airway resistance   The narrowing of the airway as a result of an exposure to an irritant material.

AL   Action level.

ALARA   Acronym for as low as reasonably achievable. A basic concept of radiation protection which specifies that radioactive discharges from nuclear plants and exposure of personnel to ionizing radiation be kept as far below regulatory limits as is reasonably achievable.

alarm set point   The selected concentration at which an instrument is set to alarm.

albumin   A protein found in nearly every animal.

albuminuria   Presence of serum albumin in the urine.

alcohol-type foam   A fire fighting foam that is effective against many water soluble and nonwater soluble materials.

algicide   A chemical agent added to water to destroy algae.

algorithm   An accepted procedure that has been developed for the purpose of solving a specific problem.

aliphatic   Organic compounds with an open chain structure as opposed to aromatic compounds with a ring structure.

aliquot   A part that is a definite fraction of the whole, such as an aliquot of a sample for analysis.

ALJ   Administrative law judge.

alkali   Any substance that is bitter in water solution, is more or less irritating or caustic to the skin, turns litmus blue, and has a pH value greater than 7.

allegation   A statement offered without proof.

allergen   An antigen that induces a response in the immune system such that subsequent exposures to the same allergen produce an allergenic reaction or specific hypersensitivity.

allergic asthma   A common form of asthma due to hypersensitivity to an allergen.

allergic contact dermatitis   An inflammatory response caused by exposure of an individual to an irritating material, with resulting edema and erythema. Initial exposure of an individual to a chemical may not cause a problem, but will result in the formation of antigens. For some people, subsequent exposure can result in allergic contact dermatitis.

allergic reaction   Abnormal response following exposure to a substance by an individual who is hypersensitive to that substance as a result of a previous exposure.

allergic rhinitis   An inflammation of the mucous membranes in the nose.

allergy   The acquired hypersensitivity of an individual to a particular substance. A hypersensitive or pathological reaction by a person to environmental factors or substances, such as pollens, foods, dust, or microorganisms, in amounts that do not affect most people. An abnormal response of a hypersensitive person to a chemical or physical stimulus.

allotrope   One of several possible forms of a substance. Carbon is an example.

alloy   A combination of two or more metals in which the atoms of one metal replace or occupy interstatial positions between the atoms of the other metal.

alopecia   Loss of hair due to toxic response to a chemical.

alpha emitter   A radioactive substance which gives off alpha particles.

alphanumeric   Consisting of alphabetical and numerical symbols. Any letter of the alphabet and a number.

alpha particle   The charged particle emitted from the nucleus of an atom that decays by alpha emission. The alpha particle has a mass and charge equal in magnitude to that of a helium nucleus (i.e., two protons and two neutrons). Exposure to alpha radiation is primarily an internal radiation hazard.

alpha radiation   *See* alpha particle.

alpha ray   *See* alpha particle.

altimeter   An instrument for determining elevation or height above a reference level.

aluminosis   A pneumoconiosis that results from the inhalation of aluminum-bearing dusts.

alveolar   Concerning the air spaces within the lungs.

alveoli   Plural of alveolus. When used in reference to the lungs, it refers to the very small dilations at the end of the bronchioles through which oxygen is taken up by the blood and carbon dioxide is released from the blood. Thus, the alveoli comprise the gas exchange region of the lungs.

A/m   Amperes per meter.

AMA   American Medical Association.

amalgam   A mixture or alloy of mercury with any of a number of other metals or alloys.

amalgamation   The alloying of metals with mercury.

ambient   The surroundings or the area encircled. Pertaining to localized conditions, such as temperature, humidity, or atmospheric pressure.

ambient air   The surrounding air or atmosphere in a given area under normal conditions. The part of the atmosphere that is external to structures and to which the public has access.

ambient air quality   Quality of the open or ambient air.

ambient air quality standard   [AAQS]   A limit of the amount of an air pollutant that is permissible in the ambient air.

ambient noise   The noise associated with a given environment and composed of the sounds from many sources. The total noise energy, or the composite of sounds from many sources in an environment.

ambient pressure   The total pressure of the environment.

ambient size-selective criteria   Measurement technique used in industrial hygiene sampling and outdoor air pollution studies to assess the potential health effects of airborne particles of different size fractions.

ambient temperature   The temperature of the medium that surrounds an object.

amblyopia   Reduced vision or dimness of vision.

AMCA   Air Moving and Control Association.

ameliorate   To ease or improve.

amended water   Water to which a wetting agent has been added to improve its ability to wet a material.

American Academy of Industrial Hygiene   [AAIH]   A professional society of board certified industrial hygienists.

American Board of Industrial Hygiene   [ABIH]   Specialty board whose objective is to improve the practice and educational standards of the profession of industrial hygiene, and that is authorized to certify qualified industrial hygienists in the discipline of industrial hygiene.

American Association of Occupational Health Nurses Inc.   [AAOHN]   An association that provides board certification in the specialty of occupational health nursing.

American Conference of Governmental Industrial Hygienists   [ACGIH]   A professional association of individuals employed by various governmental agencies and who are involved in industrial hygiene programs that are carried out or supported by governmental units.

American Industrial Hygiene Association   [AIHA]   An association of professional industrial hygienists trained in the anticipation, recognition, evaluation, and control of health hazards, and the prevention of adverse health effects among personnel in the workplace and surrounding community.

American Medical Association   [AMA]   Professional association of persons holding medical degrees or unrestricted licenses to practice medicine with the purpose of promoting the science of medicine and the betterment of public health.

American National Standards Institute   [ANSI]   A voluntary organization made up of members who coordinate, develop, and publish consensus standards for a wide variety of conditions, procedures, and devices.

American Occupational Medical Association   [AOMA]   Professional society of medical directors and plant physicians specializing in occupational medicine and surgery. The organization was established to encourage the study of problems peculiar to the practice of industrial medicine, to develop methods to conserve the health of workers, and to develop an understanding of the medical care needs of workers.

American Petroleum Institute   [API]   A trade association of major oil companies that sets standards of performance for that industry and publishes petroleum statistics.

American Public Health Association   [APHA]   An association of health professionals that develops health standards and policies.

American Society of Heating, Refrigerating, and Air Conditioning Engineers   [ASHRAE]   A professional society of heating, ventilating, refrigeration, and air conditioning engineers that carries out research programs and develops recommended practices/guidance in these areas.

American Society of Mechanical Engineers [ASME]　A professional mechanical engineers' organization that develops standards and codes appropriate for elevators, powered industrial trucks, compressed air systems, and other equipment/systems that are addressed in this phase of engineering.

American Society of Safety Engineers [ASSE]　An organization of safety personnel dedicated to the development and promotion of safety practices and standards as well as the education of members and workers.

American Society for Testing and Materials [ASTM]　An organization whose members are from business, the scientific community, governmental agencies, educational institutions, laboratories, etc., and who establish voluntary consensus standards for materials, products, systems, and services.

American Standard Code for Information Interchange [ASCII]　The most common convention for representing alphanumeric data for transmission or storage.

Americans with Disabilities Act (ADA)　Federal regulation to protect workers with disabilities from discrimination in hiring and on the job.

Ames test　A test to determine the mutagenicity of a substance using bacteria as the test medium.

AML　Acute myelogenous leukemia.

ammeter　Instrument used to measure the amount of electric current in a circuit.

amorphous　Noncrystalline and without definite form.

amosite asbestos　An asbestiform mineral of the amphibole group made up of straight brittle fibers that are light gray to pale brown in color; often referred to as brown asbestos.

amp　Ampere(s).

ampere　The meter–kilogram–second unit of electricity equal to a flow of 6.25 E^{18} electrons per second.

amphibole　One of two major groups of minerals.

amphibole asbestos　Fibrous silicates of magnesium, iron, calcium, and sodium that are generally brittle. This form of asbestos is more resistant to heat than the serpentine (chrysotile) type.

amphoteric　Material having the capacity of behaving as an acid or a base.

amplification (Biology)　The multiplication of microbiological organisms.

amplifier　A device for increasing the strength of an electric signal.

amplitude (Acoustics)　The maximum value (i.e., decibels) of a periodic varying quantity.

amt　Amount.

amu　Atomic mass unit.

anabolism　The metabolic process by which simple substances are synthesized into the complex materials of living tissue.

anaerobe　Microorganism that grows in the absence of air.

anaerobic bacteria　Bacteria that do not require free oxygen to live or that are not destroyed by its absence.

analgesic   A substance used for relieving pain.

analog   A system, such as the output of a meter, in which numerical data are represented by analogous physical magnitudes or electrical signals that vary continuously.

analyte   The substance/contaminant being analyzed for in the analytical procedure.

analytical blank   Sampling media that has been set aside for analysis but was not taken into the field. Its analytical result indicates whether the sampling media was contaminated and to what amount.

analyzer (Acoustics)   A combination of filters and a system for indicating the relative energy that is passed through the filter system. The measurement is usually interpreted as giving the distribution of energy of the applied signal as a function of frequency.

anaphylaxis   An unusual or exaggerated allergic reaction of an organism to a foreign protein or other substance following previous contact with that material.

anechoic room   A room whose boundaries (i.e., walls, ceiling, etc.) effectively absorb all the sound that is incident on their surface, thereby creating essentially a free-field condition. Also referred to as a free-field room.

anemia   A pathological deficiency of the oxygen-carrying material of the blood measured in unit volume concentrations of hemoglobin, red blood cell volume, and red blood cell number.

anemic hypoxia   A reduced oxygen supply to tissues due to impaired oxygen transport by the blood.

anemometer   A general term for instruments designed to measure the speed of the wind or the air velocity associated with ventilation systems.

aneroid barometer   A barometer which measures atmospheric pressure using one or more aneroid capsules in series.

aneroid capsule   A thin metal disc partially evacuated of air and used to measure atmospheric pressure by measuring the expansion or contraction of the capsule as the pressure changes.

anesthetic   A chemical that has a depressant effect on the central nervous system, particularly the brain, and that induces insensibility to pain.

anesthetic effect   A loss of the ability to perceive sensory stimulation. The loss of sensation with or without the loss of consciousness.

aneurysm   A pathological blood-filled dilation of a blood vessel.

angina   Any disease characterized by choking or suffocation.

angiosarcoma   A malignant growth on the inner linings of blood vessels, typically found in areas of high blood vessel concentration such as the liver. Vinyl chloride monomer is known to cause angiosarcoma of the liver.

angiospasm   The spasmodic contraction of blood vessels.

angstrom   [Å]   Unit of wavelength equivalent to 1 E^{-8} cm.

anhydrous   Without water; a substance free of water.

animal biosafety levels (Laboratory)   Facilities and procedures for ensuring appropriate levels of environmental quality, safety, and animal care for work in which experimental animals are used. There are four published levels of standards and recommended practices for use when vertebrate animals are used. These are described in the AIHA Biohazards Reference Manual.

anion   A negatively charged ion.

anneal   To subject a material to a process of heating and slow cooling in order to toughen it and reduce brittleness.

annihilation (Ionizing Radiation)   Process by which a negative electron and a positive electron combine and disappear with the emission of two gamma rays (annihilation radiation) having an energy of 0.511 MeV.

anode   Positive electrode. The electrode to which negative ions are attracted.

anodyne   *See* analgesic.

anomaly   A process, entity, organism, or part of an organism that deviates from normal.

anorexia   The lack of or loss of appetite for food.

anosmia   The absence of the sense of smell.

anoxemia   The reduction of the oxygen content of the blood to below physiologic levels.

anoxia   A condition in which there is an absence or lack of oxygen, or a reduction of oxygen, in body tissues below physiologic levels. This undersupply of oxygen reaching the tissues of the body can possibly cause permanent damage or death.

ANPRM   Advanced Notice of Proposed Rule-Making.

ANSI   American National Standards Institute.

antagonistic   The action of a substance to nullify the action of or act against another substance. Opposition in the action between similar things, as between medicines, chemicals, muscles, etc.

anthracosilicosis   A complex, chronic pneumoconiosis that is a combination of anthracosis and silicosis.

anthracosis   Usually an asymptomatic form of pneumoconiosis caused by the deposition of anthracite coal dust in the lungs.

anthrax   A highly infectious disease of ruminants (e.g., cattle, sheep, goats, etc.) that is transmissible to man by direct contact with an infected animal or by the inhalation of anthrax spores that are released into the air from contaminated ground or from the skins of infected animals.

anthropogenic   The scientific study of the origin of man, or something caused directly or indirectly by humans.

anthropometric evaluation   A study of body size and actions with the objective of improving the design of machines and tools to enable more effective use of them by humans.

anthropometry   The collection and application of body measurements as design criteria for improving the efficiency and comfort of humans in the work setting or other environment.

antibacterial   A substance with the capacity to destroy or suppress the growth or reproduction of bacteria.

antibiotic   A substance used to kill bacteria and stop the growth of disease microorganisms in the body.

antibody   Proteins in the blood that are generated as a result of a reaction to a foreign protein or polysaccharide. These neutralize the foreign proteins, and, as a result, can produce immunity against certain microorganisms or their toxins.

anticoagulant   Any substance that suppresses, delays, or prevents the coagulation of blood.

antidote   A remedy to counteract the effects of a toxic substance.

antigen   Any substance that, when introduced into the body, stimulates the production of an antibody.

antigenicity   The ability of an agent to produce a local or systemic immunologic response in a host.

antihistamine   A substance that counteracts the effect of histamine and relieves the symptoms of allergic reactions.

antiknock agent   A substance, such as an organic lead compound (e.g., TEL/TML) that can be added to motor fuel to reduce engine knocking.

antimicrobial   Agent that kills microbial growth.

antioxidant   A chemical compound that inhibits the oxidation of a material or the deterioration of a material by the process of oxidation.

antipyretic   Medication for reducing fever or inflammation.

antiseptic   Substance that prevents or inhibits the growth of microorganisms.

antitoxin   Antibodies produced within the body in response to the presence of a toxin.

antivenin   Serum used hypodermically to counteract or neutralize snake venom.

anuria   The absence of the excretion of urine from the body.

anxiety   A feeling of apprehension or fear that something unpleasant or dangerous will occur.

AOA   American Optometric Association.

APCA   Air Pollution Control Association.

APF (Respirator)   Assigned protection factor.

APHA   American Public Health Association.

aplastic anemia   A condition in which the bone marrow fails to produce an adequate supply of red blood cells.

apnea   The temporary cessation of breathing.

appeal   A formal request for the review of a lawsuit or fine and a reversal of the verdict or decree.

apprehension   A fearful or uneasy anticipation of the future or a future task to be carried out.

approved   Designation of an item that has been tested and found to be acceptable by a recognized authority and approved for use under specified conditions. Testing agencies (e.g., for IH/safety issues) include the U.S. Bureau of Mines, NIOSH, U.S. Department of Agriculture, and others.

approved equipment   Equipment that has been tested and approved by an appropriate authority (e.g., NIOSH) as safe for use in a specified hazardous atmosphere.

approved landfill   Site that has been approved by a government environmental protection agency for the disposal of hazardous wastes.

approximation   An estimate that is nearly correct.

APR   Air-purifying respirator.

a priori   A inference of an effect or outcome from a theory regarding cause.

AQCR   Air quality control region.

aquatic toxicity   The adverse effects to marine life that result from their exposure to a toxic substance.

aqueous   Pertaining to, similar to, containing, or dissolved in water.

aqueous humor   The fluid in the anterior (front) chamber of the eye.

aquifer   Underground water reservoir contained between layers of rock, sand, or gravel and capable of yielding a significant amount of groundwater to wells or springs.

arbovirus   A group of viruses that are transmitted to man by various mosquitoes and ticks.

arc   The discharge of electricity that occurs when electricity jumps between two points.

arc-welders' disease   A pneumoconiosis resulting from the inhalation of iron particles; may also be referred to as siderosis.

area sample   An environmental sample obtained at a fixed point in the workplace; used to measure properties of the workplace itself that may or may not correlate with results of individual worker (personal) samples.

argyria   Poisoning by silver or a silver salt; a prominent symptom is a permanent gray discoloration of the skin, conjunctiva, and internal organs.

arithmetic mean   The sum of values divided by the number of values.

aromatic hydrocarbon   A major group of unsaturated cyclic hydrocarbons containing one or more rings.

arraignment   The appearance of a defendant in a criminal prosecution before the court to answer the allegations made against the defendant and to enter a plea of guilty or not guilty.

arrestance   The ability of a filter to remove coarse particulate matter from air passed through it.

arrhythmia   Any irregularity in the force or rhythm of the heartbeat.

artery   A vessel through which blood passes away from the heart to capillaries in the various parts of the body.

arthralgia   Pain in a joint.

article   A manufactured item that is an individual thing in a class with its function dependent on its design.

artificial fiber   Fibers made from glass, rayon, nylon, refractory materials, or other man-made substances; also referred to as man-made mineral fibers.

artificial radioactivity   Radioactivity produced by the bombardment of a target element with nuclear particles.

artificial respiration   Any of various methods restoring normal breathing in an asphyxiated but living person, usually by rhythmic forcing of air into and out of the lungs.

asbestiform mineral   A mineral that, due to its crystalline structure and chemical composition, tends to be separated into fibers and that can be classed as a form of asbestos. The EPA defines asbestiform as a specific type of mineral fibrosity in which the fibers and fibrils possess high tensile strength and flexibility.

asbestos   Naturally occurring hydrated mineral silicates possessing a unique crystalline structure; includes the asbestiform varieties of chrysotile, crocidolite, amosite, anthophyllite, tremolite, and actinolite. The various forms of asbestos are noncombustible in air, and several can be separated into fibers.

asbestos abatement   Procedures to control the release of asbestos fibers from asbestos-containing materials.

Asbestos Analysts Registry   A listing of asbestos counters who have met the requirements of the AIHA.

asbestos bodies   Dumbbell-shaped bodies that may appear in the lungs and sputum of persons who have been exposed to asbestos. These are also called ferruginous bodies.

asbestos-containing material   [ACM]   Material that contains more than 1 percent asbestos by weight.

asbestos-containing waste material   Friable asbestos waste.

asbestos fiber (Industrial Hygiene)   An asbestos fiber that is greater than 5 micrometers in length and with a length to width ratio equal to or greater than 3 : 1.

asbestosis   A nonmalignant, progressive, irreversible lung disease characterized by diffuse fibrosis and resulting from the inhalation of asbestos fibers.

A-scale sound level   The level of sound that approximates the sensitivity of the human ear.

ASCII   American Standard Code for Information Interchange.

asepsis   The condition of being clean and free of microorganisms.

aseptic   Free from infection; sterile.

aseptic technique   The performance of a procedure or operation in a manner that prevents the introduction of septic material.

ash   The inorganic residue remaining after ignition of combustible substances.

ASHARA   Asbestos School Hazard Abatement Reauthorization Act.

ashing   The decomposition, prior to analysis, of the organic matrix constituents of a sample and sampling media.

ASHRAE   American Society of Heating, Refrigerating, and Air Conditioning Engineers.

askarel   Generic term for a group of nonflammable, synthetic, chlorinated hydrocarbons that have been used as electrical insulating material. These are also referred to as polychlorinated biphenyls.

ASME   American Society of Mechanical Engineers.

aspect ratio (Asbestos Fibers)   The ratio of fiber length to fiber width.

aspect ratio (EPA)   The ratio of the length to the width of a particle. The aspect ratio for counting structures as defined in the TEM method of asbestos sample assessment is equal to or greater than 5 : 1.

aspect ratio (OSHA)   The ratio of the length to the width of a particle; to be counted as a fiber by the PCM method, the fiber must be at least 5 micrometers in length and have an aspect ratio of at least 3 : 1.

aspergillosis   An infectious disease of the skin, lungs, and other parts of the body caused by certain fungi of the genus *Aspergillus*.

asphyxia   A high degree of respiratory distress or suffocation due to lack of oxygen. A condition due to lack of oxygen in inspired air, resulting in loss of consciousness or actual cessation of life.

asphyxiant   A substance that is capable of producing asphyxia or has the ability to deprive tissues of oxygen. Asphyxiants may be simple or chemical. The former are materials that can displace oxygen in the air (e.g., nitrogen), and the latter (e.g., carbon monoxide) render the body incapable of utilizing an adequate supply of oxygen. Both types of anoxia can potentially result in insufficient oxygen in the blood to sustain life.

asphyxiation   Suffocation as a result of being deprived of oxygen.

aspirate   The accidental passage of a liquid or solid substance into the lungs following attempted ingestion or during a vomiting sequence.

aspiration   A hazard to the lungs following the ingestion (accidental or purposeful) of a material, such as a solvent or solvent-containing product, when a small amount of the material is taken into or is aspirated into the lungs in liquid form. Aspiration can occur during ingestion or if and when the material is later vomited. Aspiration of a chemical can occur if pipetting is done by mouth.

aspiration hazard   A material being used in such a manner that it presents a danger of drawing it into the lungs. This hazard can result if vomiting occurs or is induced.

aspirator   Any device that removes liquids or gases from a space by suction.

assay   The qualitative or quantitative analysis of a substance.

ASSE   American Society of Safety Engineers.

assessment   The critical analysis, evaluation, or judgment of the status or quality of a particular condition, situation, or other subject of appraisal.

assigned protection factor   [APF]   A numerical indicator of how well a respirator can protect its wearer under optimal conditions of use. The minimum anticipated level of protection provided by each type of respirator worn in accordance with an adequate respiratory protection program. The numerical value, or assigned protection factor, is the ratio of the air contamination concentration outside a respirator to that inside the respirator. For example, an assigned protection factor of 10 means that one-tenth the workspace exposure concentration is inhaled by the wearer.

astasia   An inability to walk or stand due to muscular incoordination.

asthenia   Lack or loss of strength or energy.

asthma   Constriction of the bronchial tube muscles in response to irritation, allergy, or other stimulus, resulting in shortness of breath, wheezing, and mucus secretions.

astigmatism   The unequal curvature of the refractive surfaces of the eye resulting in a ray of light not being focused sharply on the retina but being spread over a diffuse area.

ASTM   American Society for Testing and Materials.

asymptomatic   Having a disease agent in the body but showing no outward signs of the disease; the lack of identifiable signs or symptoms.

asymptotic curve   A curve that approaches but never reaches zero.

ataxia   A failure or lack of muscular coordination.

atelectasis   Collapse of the adult lung.

atm   Atmosphere.

atmosphere   [atm]   A unit of pressure equal to the atmospheric pressure at sea level. The envelope of air surrounding the earth.

atmosphere-supplying respirator   A respiratory protective device that is designed to supply breathing air to the wearer. This type of respirator does not rely on the use of air from the work environment. The air is obtained from an independent source. A respirator of this type is classified as a supplied-air respirator or a self-contained breathing apparatus.

atmospheric pressure   The pressure exerted by the weight of the atmosphere equivalent to 14.7 pounds per square inch at sea level. Also equivalent to the pressure exerted by a column of mercury 760 millimeters high or a column of water 406.9 inches high.

atmospheric tank   A storage tank that has been designed to operate at pressures from atmospheric to 0.5 psig.

at. no.   Atomic number.

atom   The smallest part of an element that still retains the chemical properties of that element. A particle of matter that is indivisible by chemical means.

atomic absorption spectroscopy   [AAS]   Analytical method for determining the amount of a specific metal element in a sample.

atomic energy   The energy released in nuclear reactions.

atomic mass   The mass of a neutral atom of a nuclide, usually expressed in terms of atomic mass units.

atomic mass unit   [AMU]   The mass equal to 1/12 of one neutral carbon 12 atom. Equivalent to 1.6604 E^{-24} grams.

atomic number   The number of protons in the nucleus of an atom.

atomic power   The production of electricity through the use of a nuclear reactor.

atomic waste   The atomic ash produced by a splitting of nuclear fuel.

atomic weight   [at. wt.]   The sum of the number of protons and neutrons in the nucleus of an atom.

atopic dermatitis   A chronic pruritic eruption that occurs in humans. It is of unknown etiology, although allergic, hereditary, and psychogenic factors appear to be involved.

atresia   Congenital condition in which the ear canal never fully develops.

at risk   Being subject to or exposed to a substance, material, condition, or environment that may injure or result in harm to the individual.

atrophy   The wasting away or the diminution in the size of a cell, tissue, organ, or part.

attack   Onset of an illness.

attack rate   The cumulative incidence of an infection in a group observed over a period during an epidemic.

attendant (Confined Space Entry)   A trained individual outside a confined space who acts as an observer of the authorized entrants within the confined space, keeping in constant, though not necessarily continuous, communication with them, so the attendant can immediately call rescue services if needed.

attended operation   An operation which is attended at all times by a person who is sufficiently knowledgeable to act should the need arise.

attenuate   To reduce in amount.

attenuation (Acoustics)   The reduction, expressed in decibels, of the sound intensity at a designated position as compared to the sound intensity at a second position acoustically farther from the source or as a result of an intervening material.

attenuation (Ionizing Radiation)   The process by which a beam of ionizing radiation is reduced in intensity when passed through a material.

atto   [a]   Prefix designating 1 E^{-18}.

attribute   A qualitative characteristic of an individual or an item.

attrition   The wearing or grinding down of a substance by friction.

at. wt.   Atomic weight.

atypical   Unusual or abnormal.

audible range   The frequency range over which the normal ear hears sound; from about 20 to 20,000 hertz.

audible sound   Audible sound containing frequency components in the range from 20 to 20,000 hertz.

audiogram   A graph or record showing the hearing threshold of each ear of an individual as a function of frequency. The test frequencies are typically from 500 to 6000 hertz.

audiologist   A person trained in the science of hearing.

audiology   The science of hearing, particularly the study of impaired hearing that cannot be improved by medication or surgical therapy.

audiometer   An instrument for determining the hearing threshold level of an individual. This device presents a pure tone sound at a series of frequencies to which a subject responds when the sound is perceived. The result, referred to as an audiogram, is a measure of an individual's hearing acuity.

audiometric technician   An individual who is trained to perform audiometric testing.

audiometric zero   The threshold of hearing, which is equivalent to a sound pressure of 2 E^{-4} microbars.

audiometry   The testing of the sense of hearing.

audiovisual   The simultaneous stimulation of both the senses of hearing and sight.

audit   A detailed, systematic review of an occupational health and safety program to determine compliance with company policies, practices, and procedures, as well as the regulations that are applicable to the operations and work being carried out.

auditory   Pertaining to the sense of hearing.

auditory fatigue   *See* temporary threshold shift.

aural   Pertaining to the ear or hearing.

aural insert   An insert-type hearing protection device.

auricle   The part of the ear that projects from the head; also, one of the two upper chambers of the heart.

ausculation   Diagnostic monitoring done by listening to the sounds made by internal organs or any internal body part.

authority having jurisdiction   The organization, office, or individual responsible for approving equipment, an installation, or a procedure.

authorized entrant (Confined Space)   An employee who is authorized by the employer or the designee of the employer to enter a confined space.

autoclave   An apparatus for effecting sterilization using steam under pressure.

auto-ignition   The ignition of a combustible material, without initiation by a spark or flame, when the material has been raised to a temperature at which self-sustaining combustion occurs.

auto-ignition temperature   The lowest temperature at which a flammable gas-air or vapor-air mixture ignites from its own heat source or a contacted hot surface, but without the presence of a spark or flame.

automation   A general term used to indicate that an operation of any kind is partially or fully automatic.

autonomic nervous system   Nervous system in vertebrates which regulates the vital internal organs in an involuntary manner.

autopsy   The detailed examination of the body following death.

autotrophic   Description of an organism that can build its own organic constituents from carbon dioxide and inorganic salts.

auxiliary air hood   Fume hood equipped with a supply air plenum outside the hood at the top of the face opening to provide a downward-flowing air stream into the face of the hood. It is so designed to reduce the exhausting of tempered room air.

average   [avg.]   An arithmetic term indicating the value arrived at by finding the sum of a number of values and dividing that sum by the number of values.

average precipitation   The average amount of precipitation (rain or snow) that occurs over a specified period.

average temperature   The average of all temperature readings throughout the year.

averaging time   The time period over which a function is measured, yielding a time-weighted average.

aversion response   The movement of the eyelid or head to avoid exposure to bright light.

avg.   average.

avicide   A product used to kill birds or as a bird repellant.

Avogadro's number   The number of molecules in a gram molecular weight (mole) of a substance. It is equivalent to 6.023 E^{23} and is one of the fundamental constants of chemistry.

A-weighted network   Weighting network that is presented on sound level meters and octave band analyzers which mimic the human ear's response to sound.

A-weighted sound level   [dBA]   The sound level determined by employing the A scale of a sound level meter, or other noise survey meter equipped with this weighting network.

AWS   American Welding Society.

axiomatic   Relating to a self-evident or universally recognized truth.

axon   The core of a nerve fiber that generally conducts impulses away from the nerve cell.

azeotrope   A liquid mixture that has a constant boiling point different from that of its constituents and that distills without change of composition.

azotemia   An excess of urea or other nitrogenous bodies in the blood.

B

backdraft (Fire)   An explosion that occurs when oxygen suddenly gains access to a smoldering fire.

backdraft (Ventilation)   A condition in which negative pressure in a building can prevent the effective discharge from exhausts stacks/vents and can draw emissions back into the building.

backfill   Material used to refill a ditch or other excavation, or the process of doing it.

backflow   The flow of water or other liquid into the distribution pipes of a potable water supply system from any source other than the source of the potable water.

backflow connection   A plumbing arrangement in which backflow can occur.

backflow preventer   A device or means that prevents the backflow of a liquid into the potable water supply.

background contamination   Substance in the air that is typically present from sources other than those used to assess an exposure.

background level (Air Pollution)   The amount of a pollutant that is present in the ambient air due to natural sources.

background noise   Noise from sources other than the direct airborne sound emitted from the source being evaluated or measured.

background radiation   The ionizing radiation coming from sources other than the radiation source (material or X-ray producing equipment) being measured.

backscatter   The scattering of ionizing radiation in a reverse direction to that of the primary beam.

backsiphonage   Backflow resulting from negative pressure in the distributing pipes of a potable water supply.

BACT   Best available control technology.

bacteria   Any of numerous unicellular microorganisms of the class Schizomycetes, occurring in a wide variety of forms, existing either as free-living organisms or as parasites, and having a wide range of biochemical and often parasitic properties.

bacterial food poisoning   An illness due to the ingestion of food containing a toxin that has resulted from bacterial contamination and growth in the food.

bactericide   A substance that destroys bacteria. Also referred to as a bactericidal agent.

bacteriophage   A submicroscopic organism, usually viral, that destroys bacteria.

bacteriostatic agent   A material that stops the growth and multiplication of bacteria, but does not necessarily kill them.

baffle (Acoustics)   A partition used to increase the path length of a sound from its source of generation to a receptor, thereby reducing the sound level reaching the receptor.

bagasse   The waste from sugar cane after the sugar has been extracted.

bagassosis   A lung disease, or pneumoconiosis, produced as a result of the inhalation of the dust of bagasse, the waste of sugar cane after the sugar has been extracted. Bagasse, which is moist and recently ground, is not believed to cause this disease.

bag-house   A fabric type dust collector that is typically a steel structure containing a cloth filtering system for removing particulates from the air drawn through it by a fan. The present term for this particulate control system is a fabric collector.

bakeout   The procedure of overheating a new building or space for several days before occupancy, then flushing it out with 100 percent outside air to remove contaminants that may contribute to poor indoor air quality.

balancing by dampers   Method for designing a local exhaust system and its ductwork using adjustable dampers to distribute airflow after system installation.

balancing by static pressure   Method for designing a local exhaust system and its ductwork by selecting the duct diameters that generate the static pressure to distribute the desired airflow throughout the system without the use of dampers.

ballast (Present Day)   Water taken on board a ship to permit proper trim angle and draft for the vessel in the water. Typically salt water is used, but iron or concrete may be used as well.

band (Acoustics)   A segment of the frequency spectrum of noise.

band level   *See* band-pressure level.

band-pass filter   A filter with a single transmission band extending from lower to upper cutoff frequencies.

band-pressure level   The sound pressure level for sounds contained within a restricted (specified) frequency band.

BaP   Benzo(a)pyrene.

BAQA   Bulk Asbestos Quality Assurance Program.

bar   Barometer.

bar   Unit of pressure equal to 1 E^5 N/m ^{2} or 0.98697 standard atmospheres.

bar diagram   A graphic technique for presenting data organized in a manner such that each observation can fall into one and only one category of the variable.

baritosis   A form of pneumoconiosis resulting from the inhalation of barium sulfate or other barium compound.

barograph   A continuous recording barometer.

barometer   An instrument for measuring the pressure of the atmosphere.

barometric pressure   The pressure at a given temperature and altitude due to the pressure of the atmosphere.

barotrauma   Injury caused to the wall of the eustacian tube and the ear drum as a result of the difference in pressure between that of the atmosphere and that in the middle ear.

barrel   [bbl]   A unit of volume for petroleum products equivalent to 42 U.S. gallons at 60° F.

barrier cream   Cream that is applied to the skin to provide protection against contact with various chemicals and solvents.

barrier guards   Devices that enclose the danger zone of a machine before the machine can be turned on.

basal cell carcinoma   A skin tumor which usually appears as a flaky lump or nodule on the head, neck, hands, or trunk of the body.

basal metabolism   The amount of energy required by the body when at rest.

base   A substance which produces hydroxyl ions in water and reacts with acids to form salts and water.

baseline audiogram   An audiogram against which future audiograms are compared to determine hearing threshold shifts at various test frequencies. This is typically the first audiogram taken during employment with the present employer.

baseline data   Data that describes the magnitude and range of exposures for a homogeneous exposure group and stressor (e.g., an airborne contaminant, physical agent, etc.).

basophil   A cell, especially a white blood cell, having granules that have an affinity for basic dyes.

BAT   Best available technology.

batch   A specific quantity of material that is processed, treated, or used in one operation.

batch process   A process in which the charge or feedstock is added intermittently in definite portions or batches as opposed to continuously.

BATEA   Best available technology economically available.

battery life (Instrument)   The time period over which the battery of an instrument will provide sufficient power for uninterrupted operation of the device.

bauxite   Aluminum ore. A natural aggregate of aluminum-bearing minerals in which the aluminum occurs as hydrated oxides.

bauxite pneumoconiosis   A rapidly progressive pneumoconiosis leading to extreme pulmonary emphysema as a result of the inhalation of bauxite fumes containing fine particles of alumina and silica. Also called Shaver's disease.

bbl   Barrel.

BC (Hearing)   Bone and tissue conduction.

bcf   Billion cubic feet.

BCG vaccine   A preparation used as an active immunizing agent against tuberculosis.

BCSP   Board of Certified Safety Professions.

beam   Unidirectional ray of ionizing or non-ionizing radiation.

bearing wall   A wall that supports a vertical load, such as a floor or roof, in addition to its own weight.

beat elbow   Bursitis of the elbow joint that can result from the use of heavy vibrating tools.

beat knee   Bursitis of the knee joint as a result of friction or vibration.

becquerel   [Bq]   A unit expressing the rate of radioactive disintegration. One becquerel is equal to one radioactive disintegration per second. There are 3.7 E^{10} becquerels per curie of radioactivity.

Beer-Lambert Law   Physical law that states that the absorptivity of a substance is a constant with respect to changes in concentration.

behavioral science   The study of natural phenomena related to and the laws governing the behavior of man and the lower animals as social beings.

BEI   Biological Exposure Indices.

BEIR Committee   Biological Effects of Ionizing Radiation Committee of the National Academy of Sciences. This committee reports on the health effects of ionizing radiation.

bel   The logarithm to the base 10 of the ratio of two levels of power.

Belding-Hatch Index   A method for determining the heat stress experienced by an average man for various types of activities. It is also referred to as the Heat-Stress Index.

benchmark (Epidemiology)   A measurement taken at the outset of a series of measurements of the same variable.

bends   Pain in the limbs and abdomen occurring as a result of a rapid reduction of air pressure. Also referred to as caisson worker's disease or decompression sickness. *See* decompression sickness.

benefit   The advantage or improvement realized as a result of an action taken.

benefit-cost ratio   The ratio of the present value of measurable benefits to the costs of having taken the action as well as those that would have been incurred had the action not been taken.

benign   Not malignant or recurring. Typically used to describe tumors that grow in size but do not spread throughout the body (i.e., do not metastasize or invade tissue).

benign tumor   A tumor that lacks the properties of invasion and metastasis and that is usually surrounded by a fibrous capsule.

benzene ring   *See* aromatic hydrocarbon.

beryl   A silicate of beryllium and aluminum that is considered a carcinogen.

berylliosis   An occupational disease, usually of the lungs, resulting from exposure to beryllium fumes or dust, and characterized by nodules of granulation tissue or a tumor-like mass as a result of an inflammatory process. A result of chronic inhalation of beryllium containing dust.

best available technology (Air Pollution)   A level of control that is the most effective at the time the control is being considered.

best available technology economically achievable (Air Pollution)   The level of emission control to be achieved that is the best technology in use.

beta decay   The process by which a radionuclide decays by beta emission.

beta particle   A charged particle emitted from the nucleus of an atom, with a mass and charge equal in magnitude to that of an electron.

BeV   Billion electron volts (1 E^9 eV).

BGT   Black globe temperature.

bhp   Brake horsepower.

biannual   Happening twice each year.

bias   A systematic error that contributes to the difference between the mean of a set of measurements and the true value. The tendency of an estimate to deviate in one direction from a true value. The monitoring method concentration minus the actual concentration, divided by the actual concentration times 100, is the percent bias.

biennial   Happening every second year.

bill of lading   A document listing the contents and acknowledging receipt of goods for shipment.

bimetallic thermometer   A thermometer which consists of two different metal strips that are brazed together. The differences in expansion of the metal strips, due to a temperature change, is used to provide an indication of temperature.

bimodal distribution   A distribution of results/observations with two regions of high frequency separated by a region of low frequency.

binary   Numbering system using only two symbols, i.e., one and zero. In a computer, the one is represented by an electric charge being on, and the zero by the absence of such a charge.

binding energy   The energy, in electron volts, that is required to hold the neutrons and protons of an atomic nucleus together. Since protons are positively charged, they exert strong repellent forces, and a significant amount of energy is required to hold them together. This is equivalent to the binding energy of the element.

binomial distribution A distribution of data/results describing probabilities of the outcome of trials that can have one or two mutually exclusive results (e.g., exposure above or below a PEL).

bioaccumulation A process in which a toxic substance collects or accumulates in the body or the environment, and, as a consequence, poses a risk to human health or the environment.

bioaerosol Airborne particulates that are or were derived from living organisms, including microorganisms as well as fragments, toxins, and waste products of all varieties of living things.

bioassay (General) A determination of the concentration of a substance in the body by the analysis of urine, blood, feces, bone, or tissue; or a measure of the change that has resulted in the body as a result of an exposure to a substance or of a metabolite that is a result of the body's having absorbed the substance. The use of the living organism to measure the amount of a substance that has been absorbed.

bioassay (Ionizing Radiation) The measurement of the radioactive materials deposited within or excreted from the body. May include whole body counting or the analysis of urine, feces, or other specimens.

bioaugmentation The use of bacteria to reduce nitrogen levels through biological digestion in effluent waters.

bioavailability The amount of a chemical that becomes available to the target organ/tissue after the material has entered the body.

biochemical oxidation *See* biological oxidation.

biochemical oxygen demand [BOD] A measure of the oxidizable components in water. The dissolved oxygen required to decompose organic matter by biological oxidation in a liquid.

biochemistry The study of the chemistry of biological substances and processes.

biocide A substance that is capable of destroying living organisms.

bioconversion The use of living organisms to transform organic waste into usable material.

biodegradable Capable of being broken down by the action of living things. Decomposition of a material through the action of microorganisms.

biodegradable material Organic waste material that can be broken down into basic elements by the action of microorganisms.

bioeffluents The chemicals emitted by people, such as carbon dioxide, body odors, etc.

bioengineering The design of equipment, structures, or work stations to fit the body characteristics of people who will work with the equipment or at the work locations.

biofilm A community of microorganisms in wastewater that attaches itself to surfaces in thick slimy layers.

biogenesis The doctrine that living organisms develop only from other living organisms and not from nonliving matter.

biohazard area An area in which work has been or is being performed with biohazards (biohazardous agents or materials).

biohazardous agent   One that is biological in nature, capable of self-replication, and has the capacity to produce deleterious effects upon other biological organisms, particularly humans.

biohazards   Synonym for biological hazards, which are biological organisms and products of biological organisms or plants that may pose a health risk to workers and others. The USPHS and the USDA have identified five classes of biohazardous agents.

biohazard warning sign   A sign used as an administrative exposure control practice for alerting personnel who may enter an area where biohazards are present so that they may take precautionary exposure control measures. Traffic in such areas can be restricted by posting such signs where they can be easily seen. These signs should only be used for the purpose of signifying the actual or potential presence of biohazards.

biokinetics   The absorption, distribution, storage, biotransformation, and elimination of a toxicant as a function of time.

biologic agents   Biologic organisms which cause infections or a disease, such as anthrax spores which cause anthrax, an occupational disease.

biological agents: Class 1   Agents of no or minimal hazard under ordinary conditions of handling.

biological agents: Class 2   Agents of ordinary potential hazard. Includes agents which may produce disease of varying degrees of severity from accidental innoculation or injection or other means of cutaneous penetration, but which are contained by ordinary laboratory techniques.

biological agents: Class 3   Agents involving special hazards or agents derived from outside the U.S. which require a federal permit for importation unless they are specified for higher classification. This class includes pathogens which require special conditions for containment.

biological agents: Class 4   Agents that require the most stringent conditions for their containment because they are extremely hazardous to laboratory personnel or may cause serious epidemic disease.

biological agents: Class 5   Foreign animal pathogens that are excluded from the U.S. by law and whose entry is restricted by USDA administrative policy.

biological control   Using other than chemical means for pest control. This may include sterilization, inhibitory hormones, or predatory organisms.

Biological Exposure Index   [BEI]   Reference value that represents the level of a determinant, such as a metabolite, which is most likely to be observed in specimens collected from a healthy worker who has been exposed to the same extent as a worker with inhalation exposure at the TLV.

biological monitoring   The measurement of the absorption of an environmental chemical in the worker by analysis of a biological specimen (e.g., blood, urine, etc.) for the chemical agent, its metabolite, some specific effect on the worker, or other appropriate indicator.

biological oxidation   The process by which microorganisms decompose complex organic materials. Also called biochemical oxidation.

biological safety cabinet   A cabinet for providing an uncontaminated area for microbiological work and for protecting laboratory workers from particulates and aerosols generated by microbiological manipulations. There are basically three main classes of these cabinets: Class I, II, and III.

biological sciences The study of living organisms including heredity, diversity, reproduction, development, structure, and function of cells, organisms, and population, with emphasis on human biology.

biological waste The waste derived from living organisms.

biologic half-life [T_b] The time required for a given species, organ, or tissue to eliminate one-half of a substance which it has taken in.

biologic test A measurement taken from biological media to determine the presence of a specific material or metabolite, or some other measurable effect on a worker that is a result of an exposure to a specific substance.

biology The study of life and life processes, including the study of structure, functioning, growth, origin, evolution, and distribution of living organisms.

biomarker A measurable biologic characteristic which has a definable relation to prior exposure to a substance.

biomass The entire assemblage of living organisms, both animal and vegetable, of a particular region considered collectively. Plant materials and animal waste used as a direct source of fuel.

biome A community of living organisms of a single major ecological region.

biomechanics A subdiscipline of ergonomics involving the study of the human body as a working system and the application of mechanical laws to it.

biomonitoring *See* biosampling.

bionics The application of biological principles to the study and design of engineering, especially electronic systems.

biophysics Science dealing with the application of physical methods and theories to biological problems/effects, such as the interaction of radiofrequency energies with living systems.

biopsy The removal and examination of tissue, cells, or fluids from a living body for examination.

biorefractory A substance that resists biodegradation and tends to persist and accumulate in the environment.

bioremediation The use of biological organisms to correct or remediate an environmental problem.

biorhythm An inherent rhythm that appears to control or initiate various biological processes.

biosafety levels Guidance provided by the U.S. Department of Health and Human Services in its publication, "Biosafety in Microbiological and Biomedical Laboratories." It addresses three factors: properly designed facility, appropriate equipment, and appropriate procedures for the work to be carried out. Four levels of biosafety guidance for such laboratories are provided.

biosampling The collection of samples (e.g., air, surface wipes, settling plates, etc.) to identify and quantify the presence of bioaerosols in the work environment.

biosolids Nutrient-rich organic material from waste treatment plants.

biosphere   That portion of the earth and its atmosphere that can support life.

biostatistics   The application of statistical methods to the analysis of biological data.

biosynthesis   The production of complex substances from simple ones by or with living organisms.

biota   The animal and plant life of a particular region considered as a total ecological entity.

biotechnology   The engineering and biological study of relationships between man and machines.

biotic agent   Microorganisms and parasites which act on the skin and body to produce disease.

biotransformation   The series of chemical alterations of a compound which occur within the body.

bipolar   Having two poles.

birth defect   Any inherited disorder or abnormality that affects newborn mammals.

bite   The nip point between two inrunning rolls of a machine, such as a calendar (roll) mill.

black body (Heat Stress)   A hypothetical ideal body which absorbs all incident radiation, independent of wavelength and direction.

black globe thermometer   Typically a 6-inch hollow, thin-walled, copper sphere painted flat black, with an ordinary thermometer placed into the globe at the center.

black light   Ultraviolet light at a wavelength between 3000 and 4000 angstroms (0.3 to 0.4 micrometers). Ultraviolet energy in this region of the electromagnetic spectrum is responsible for the pigmentation that results following exposure to ultraviolet light.

black lung disease   A pneumoconiosis resulting from the inhalation of coal dust. *See* coal miner's pneumoconiosis.

blackwater   Water from toilet usage that contains human body wastes.

blank   Unexposed sample media used in the correction of background contamination of sample results or in analyte recovery studies. They are employed to help in preventing errors in the analytical method development and identification and quantitation of analytes in field samples.

blank (Process)   *See* blind.

blanking (Process)   *See* blinding.

blank QA (spiked) sample   Sampling media spiked by quality assurance personnel with selected compounds at known amounts for submitting to the analytical laboratory along with regular samples to determine analyte recovery effectiveness, possible effects of sample storage/shipment, etc.

blast-gate   A sliding plate that is used to control the volume of air exhausted through a branch duct of a ventilation system, thereby distributing airflow through other branches of the system. This term is synonymous with damper.

blasting agent   A material which has been tested for blasting cap sensitivity, thermal stability, and fire characteristics, and found to be insensitive enough that there is very little probability of accidental initiation of an explosion or transition from deflagration to detonation.

blastomycosis   Term for any infection caused by a yeastlike organism.

blepharism   Spasm of the eyelids resulting in continuous blinking.

BLEVE   Boiling liquid expanding vapor explosion.

blind   Typically, a metal plate that serves as an absolute means to seal off a pipe, line, or duct from another section of the process. It completely covers the bore of the pipe, line, or duct, and is capable of withstanding the maximum pressure present with no leakage beyond the plate.

blinding   The procedure followed for the absolute closure of a pipe, line, or duct with a metal plate (blind) that can withstand the pressure in the pipe, line, or duct and prevent leakage beyond the plate.

blind sample   Samples that are prepared by someone other than the analyst who will analyze samples, and that are submitted to the laboratory, along with regular field samples, as an independent check on laboratory performance and the accuracy and precision of the analytical method.

blind study   A study in which observers and/or subjects are kept ignorant of the group to which subjects are assigned.

BLL   Blood lead level.

block-flow diagram   An input-output representation using block-like figures and lines to illustrate a process or system.

blood (Bloodborne Pathogens)   Human blood, human blood components, and products made from human blood.

bloodborne pathogen   Pathogenic microorganisms that may be present in human blood and that can cause disease in humans.

blood-brain barrier   A barrier that hinders the passage of many toxins from the central nervous system capillaries into the brain.

blood count   The determination of the number of various cells (white blood cells, red blood cells, etc.) in the blood.

blood dyscrasia   Any persistent change from normal in one or more components of blood.

blood level   The concentration of a material, such as lead, in the blood. Typically reported as micrograms per 100 gm of blood or micrograms per 100 mL (i.e., deciliter) of blood.

blood poisoning   *See* toxemia.

blood pressure   The pressure of the blood on the walls of the arteries, dependent on the energy of the heart action, the elasticity of the walls of the arteries, and the volume and viscosity of the blood.

blood urea nitrogen   The urea concentration in the serum or plasma. An indicator of kidney function.

BLS   U.S. Bureau of Labor Statistics.

blow (Ventilation)   The distance an air stream travels from an outlet to a position at which air motion along the axis is reduced to a velocity of 50 fpm. Also referred to as throw.

blower   A fan.

blowout   An uncontrolled flow of gas, oil, or other well fluids into the atmosphere.

BNA   Bureau of National Affairs.

Board of Certified Safety Professionals   [BCSP]   A board which establishes minimum academic and experience requirements necessary for one to qualify as a certified safety professional. The board conducts examinations for certification, maintains a listing of all those certified, issues certificates to those qualified to be certified, and establishes requirements for maintenance of certification.

BOCA   Building Officials and Code Administrators.

BOD   Biochemical oxygen demand.

bodily injury   Injury to a human as opposed to damage to property.

body burden (Toxicology)   The total amount of a chemical retained in the body.

body burden (Ionizing Radiation)   The total amount of a radioactive material retained in the body or is present in the body at a given time. The maximum amount of any radioisotope in the body that is not to be exceeded at any time, so that the dose derived from that amount of activity will not exceed the established dose limit.

body fluids   Fluids produced in the body, such as blood, urine, sweat, etc.

BOHS   British Occupational Health Society.

boilermaker's disease   The name for the very high incidence of hearing loss among this working population.

boiling liquid expanding vapor explosion   [BLEVE]   A violent rupture of a pressure vessel containing saturated liquid/vapor at a temperature well above its normal boiling point.

boiling point   The temperature at which a liquid boils, or the temperature at which the vapor pressure of a liquid is equal to the pressure of the atmosphere at the surface of the liquid.

boiling water reactor   [BWR]   Nuclear reactor in which water is used as the coolant and neutron moderator.

bolometer   An instrument which measures radiant heat by correlating the radiation-induced change in electrical resistance of a blackened metal foil with the amount of radiation absorbed.

BOM   Bureau of Mines.

bonding   A procedure for eliminating the difference in potential (i.e., electrical) between objects by connecting the objects together (i.e., metal to metal) with an appropriate wire conductor.

bone conduction   [BC]   The transmission of sounds through the bone structure of the head.

bone conduction test   A type of hearing test that is conducted to determine the nerve-carrying capacity of the cochlea and the auditory nerve.

bone marrow   The soft material which fills the cavities of most bones. Bone marrow manufactures most of the formed elements (e.g., cells) of the blood.

bone-seeker   A radioactive substance which preferentially lodges in the bone tissue when introduced into the body. Strontium-90 is considered a bone-seeker.

boroscope   An optical device that enables the visual inspection of the inside of ductwork or other inaccessible enclosures.

botulism   A type of food poisoning caused by a neurotoxin that may be present in improperly canned or preserved foods.

Bourdon tube   A closed, curved, flexible tube with an elliptical cross-section which responds to changes in barometric pressure and provides a measurement of that parameter.

Boyle's law   A physical law that states the volume of a gas varies inversely with the pressure if the temperature remains constant. This relationship is strictly true only for perfect gases, but its application is generally satisfactory, except when pressures are very high or temperatures are approaching the liquefaction point.

bp   Boiling point.

BP   Blood pressure.

Bq   Becquerel.

brackish water   Water that contains low concentrations of soluble salts.

bradycardia   Slowness of the heartbeat as evidenced by slowing of the pulse rate to less than 60.

brake horsepower   The horsepower actually required to drive a fan. Includes the energy losses in the fan and can only be determined by testing the fan.

branch (Ventilation)   A duct or pipe connecting an exhaust hood to a main or submain.

branch duct entry   The point in a ventilation system where a branch or secondary duct joins a main duct.

branch of greatest resistance   The path from a hood or duct opening to the fan and exhaust stack in a ventilation system which causes the most pressure loss.

brass   An alloy of copper and zinc which may contain a small amount of lead.

brass-founders ague   Metal fume fever that may occur in workers in brass foundries.

brattice   A partition that is placed in an underground mine to control the flow of ventilation. Often made of heavy cloth such as canvas, or of plywood.

breach   Gap, tear, or split in an enclosure or structure.

breakthrough (Sampling)   Elution of the substance being sampled from the exit end of a sorbent bed during the process of sampling air. Breakthrough is considered to occur when more than 25 percent of the contaminant found on the front section of a solid sorbent sampling tube is detected in the back section of the tube as a result of elution of the substance being sampled. Contaminant can be released from the sorbent during the process of sampling the air. Breakthrough is also noted if desorption or inefficient retention of an absorbed substance occurs.

breakthrough time (Personal Protection)   The time elapsed between initial contact of a chemical with the outside surface of a protective clothing material and the detection of the chemical at the inside surface of the material.

breath analysis   A method of determining the presence of a chemical, such as benzene or another volatile solvent, in the exhaled breath, for determining whether exposure has occurred.

breathing air   Air that equals or exceeds Grade D specifications for gaseous air in accordance with CGA G-7.1-89, and that does not present a health hazard to anyone breathing the air.

breathing zone (IAQ)   The area of a room/office in which occupants breathe as they sit, stand, or lie down.

breathing zone (IH)   Breathing area of a worker. The area or zone in the vicinity of a worker from which air is inspired. It is generally considered to be within a radius of eight to ten inches from the nose.

breathing zone sample   An air sample collected in the breathing area of a worker to assess exposure to an airborne contaminant.

breeching   The passage between a furnace and the stack that carries the products of combustion.

breeder reactor   A nuclear reactor that produces more nuclear fuel than it consumes.

bremsstrahlung   Secondary photon radiation (ionizing) associated with the deceleration of charged particles passing through matter. The term means breaking radiation, and results when high speed electrons interact with the nuclei of the absorbing substance, producing X-rays.

BRI   Building related illness.

bridging encapsulant   A material, generally in a liquid form, that is employed to seal the surface of an asbestos-containing material or other fibrous product in order to prevent the release of fibers.

brightness   The visual sensation of the luminous intensity of a light source.

British thermal unit   [Btu]   The amount or quantity of heat required to raise the temperature of one pound of water one degree Fahrenheit at 39.2 degrees Fahrenheit.

broadband noise   Noise with components extending over a wide frequency range.

bromism   Chronic poisoning as a result of excessive, prolonged exposure to bromine or the use of bromides.

bronchi   The primary branches of the trachea.

bronchial asthma   A chronic respiratory disease marked by recurrent attacks of dyspnea with wheezing due to spasmodic contraction of the bronchi.

bronchial tree   Refers to the branched structure of the air passages in the lungs.

bronchial tubes   Branches of the trachea.

bronchiectasis   Chronic dilation of the bronchi with spasmodic coughing and production of phlegm.

bronchiogenic   Originating in the bronchi.

bronchioles   The narrowest of the tubes which carry air into and out of the lungs.

bronchitis   Inflammation of the bronchi or bronchial tubes.

bronchoconstrictor   An agent that causes narrowing of the lumina of the air passages of the lungs.

bronchodilator   An agent that causes an increase in the diameter of a bronchus or bronchial tubes.

bronchopneumonia   Term indicating inflammation of the lungs, usually beginning in the terminal bronchioles, followed by their becoming clogged.

bronchoscopy   The examination of the bronchi through a bronchoscope.

bronchospasm   Spasmodic contractions of the bronchial muscle.

bronchus   Any of the larger air passages of the lungs.

brownfields   Abandoned or former industrial buildings or sites.

Brownian motion   The random movement of minute particles when they are suspended in a fluid (e.g., air).

brown lung   *See* byssinosis.

brucellosis   A generalized infection in man, resulting from contact with infected animals or consumption of infected meat or milk. Also referred to as undulant fever.

bruise   An injury to the skin and sometimes to the superficial muscles, characterized by discoloration under the skin, discomfort, and a dull persistent ache.

B.S. and W.   Bottom solids and water (storage tank bottoms).

BETX   Acronym for benzene, toluene, ethyl benzene, and xylene, which are the major aromatic components in gasoline and some other petroleum products.

Btu   British thermal unit.

Btuh   British thermal units per hour.

bubble meter   A burette or other similar volumetric device that can be used with a soap solution to form a bubble for calibrating a sampling device, such as a pump, by timing the period it takes for the bubble to traverse a specific volume and using this data to calculate its flow rate. This method is considered a primary calibration method. Also referred to as a soap-film or soap-bubble flowmeter.

bubbler   A device used to collect air contaminants by bubbling sampled air through a liquid media (e.g., absorbent) contained in the bubbler. The sampling tube of the bubbler typically has a glass frit at the end which is immersed in the collecting solution or sampling media.

buddy system   A system of organizing employees into work groups in such a manner that each employee of a work group is designated to be observed by another person in the work group. A procedure of pairing two individuals for their mutual aid or protection.

buffer   A substance that reduces the change in pH that would otherwise be produced if acids or bases are added to a solution.

building codes   Laws that set the minimum requirements for design and construction of buildings and structures in order to protect the health and safety of society. Many BOCA standards become law.

building envelope   Elements of the building, including all external building materials, windows, and walls that enclose the internal space.

Building Officials and Code Administrators   [BOCA]   A widely accepted building code.

building-related illness   [BRI]   A diagnosable illness related to poor indoor air quality, the symptoms of which can be identified and whose cause can be directly attributed to airborne building pollutants. Specific medical conditions of known etiology that can often be documented by physical signs and laboratory findings.

building services   The electrical and mechanical systems which provide power, environmental controls, and conveniences for the practical use of a building or structure.

bulk air sample   The sampling of a larger than normal volume of air through a sampling media for the purpose of determining the presence (i.e., qualitative determination) of a substance in the air rather than sampling to determine its air concentration.

Bulk Asbestos Quality Assurance Program (AIHA)   Provides bulk asbestos samples to laboratories as an external quality assurance check on the laboratory's ability to characterize bulk materials.

bulk density   The mass per unit volume of a solid material as it is normally packed.

bulk sample   The collection of a sample of the material (e.g., solvent, settled dust, bulk insulation, etc.) that is the source of the contaminant of concern. The bulk sample is analyzed to determine the presence of a component (e.g., benzene) and the amount of component of concern in the bulk sample (e.g., asbestos, lead, etc.), and to assure that the analytical method is appropriate for detecting the contaminant of concern, and for other reasons.

bullae   Bladder or sac containing liquid, such as occurs when lungs become emphysematous.

BUN   Blood urea nitrogen.

bundle (EPA: Asbestos)   A structure composed of three or more fibers in a parallel arrangement with each fiber closer than one fiber diameter.

bundle (NIOSH)   A compact arrangement of parallel fibers in which separate fibers or fibrils may be visible only at the ends of the bundle.

bung   A stopper for the hole through which a drum or barrel is filled or emptied. The hole itself is referred to as the bunghole.

buoyancy   The apparent loss of weight of an object immersed in a fluid.

burden of proof   The responsibility of proving an issue.

Bureau of Labor Statistics   Governmental (federal) body that compiles accident data on all industries by class.

Bureau of Mines   A research and fact-finding agency in the U.S. Department of the Interior with the goal of stimulating private industry to produce the country's mineral needs in ways that protect workers and the public interest.

burn   An injury resulting from exposure to fire, heat or chemicals. The classes of burns include first degree, which are characterized by redness and heat, accompanied by pain and itching; second degree, which are very painful and involve deep portions of the epidermis and the upper layer of the dermis; and third degree, which result in the loss of skin and subcutaneous tissue and are characterized by a pearly white or charred appearance.

burning rate   The time it takes a specified sample of a solid material to burn a designated distance.

burnout   A phenomenon that may occur in workers after years of high-quality performance when the individual suddenly seems unable to perform work. It is manifested by a decrease in efficiency and initiative, a diminished interest in work, and an inability to maintain work performance.

bursa   The sac or saclike cavity filled with a viscid fluid and situated at a place in the tissue where friction would otherwise develop.

bursitis   Inflammation of connective tissue around bone joints.

B-Weighted sound level   [dBB]   The sound level as determined on the B scale of a sound level meter or other noise survey meter with weighting networks.

BWR   Boiling water reactor.

bypass fume hood   A laboratory fume hood constructed such that, as the sash is closed, air bypasses the hood face via an opening that is typically located above the sash, thereby providing a reasonably constant velocity of air entering the hood face.

byproduct material   Any radioactive material obtained during the production or use of fissionable material. Includes radioisotopes obtained during the production or use of source or fissionable materials. Includes radioisotopes produced in nuclear reactors, but excludes source or fissionable material.

byssinosis   A form of pneumoconiosis, characterized by shortness of breath and chest tightness, resulting from the inhalation of cotton dust as well as hemp or flax dust. Also called cotton-mill fever and brown lung when cotton dust has been the source of exposure.

C

c   Centi, 1 E $^{-2}$.

C   Degrees Celsius, centigrade.

ca   Circa, about.

CAA   Clean Air Act.

CAAA   Clean Air Act Amendments.

cachexia   General ill health and malnutrition.

CAD   Computer aided design.

CAER   Community Awareness and Emergency Response Program.

CAGI   Compressed Air and Gas Institute.

caisson disease   *See* decompression sickness.

cal   calorie.

calcicosis   Morbid condition of the lungs due to chronic inhalation of marble dust.

calendar quarter   Any three-month period.

calcining   Exposure of an inorganic compound or mineral to a high temperature in order to alter its chemical form and drive off a substance which was originally a part of the compound.

calibrate (Instrument) The adjustment or standardization of a measuring instrument. To adjust the span or gain of an instrument so that it indicates the actual concentration of a specific substance or mixture which is present at the sensor.

calibration (Instrument) A determination of the variation of an instrument's response from a standard, and a determination of appropriate correction factors or the adjustment in the response of the device to indicate the true value. The procedure followed to determine the adjustment needed for an instrument to indicate the proper response.

calibration gas A gas of accurately known concentration which is used as a comparative standard in determining instrument performance and to adjust the instrument to indicate the true concentration.

calorie [cal] The quantity of heat required to raise the temperature of 1 gram of water 1 degree Celsius.

calorimeter A device for measuring the energy absorbed from a source of electromagnetic radiation.

CAM Continuous air monitoring.

canal caps A type of personal hearing protection which blocks noise from entering the external ear canals by placing a tight fitting cap over them.

cancer A malignant tumor anywhere in the body. A disease in which rapidly multiplying cells grow in the body, interfering with its natural functions.

candela [cd] A unit of luminous intensity equivalent to one lumen per square foot. Formerly called the candle.

candle A unit of luminous intensity.

candles per square meter [cd/m^2] Metric unit of luminance.

candlepower [cp] The unit of light intensity or brightness of a standard candle at its source.

canister (Respirator) A container filled with a sorbent, and possibly catalysts, used for removing contaminants (e.g., gases or vapors) from air being inspired through the device.

cannula (Nasal) A device consisting of two short tubes that can be inserted into the nostrils for administering oxygen or other therapeutic gases.

canopy hood A one- or two-sided hood which is positioned above an operation that typically involves heating, to receive and remove the hot air and contaminants that are released and rise and enter the hood.

CAOHC Council for Accreditation in Occupational Hearing Conservation.

CAP Chemical accident prevention.

capillary Pertaining to or resembling a hair with a very small internal diameter. One of the minute blood vessels that connect the arteries and veins.

capillary action The attraction between molecules, such as the rise of a liquid in a small diameter tube, or the wetting of a solid by a liquid.

capture gamma ray A high-energy gamma ray that is emitted when the nucleus of an atom captures a neutron and becomes intensely excited.

capture velocity The air velocity at any point in front of a hood or at the hood opening that is necessary to overcome opposing air currents and capture contaminated air at that point by causing it to flow into the hood. Sometimes referred to as control velocity.

carbon dating The use of radioactive carbon (^{14}C) to estimate the age of organic materials.

carboxyhemoglobin A compound formed between hemoglobin and carbon monoxide as a result of exposure to carbon monoxide. In this form hemoglobin is not available to carry oxygen to cells.

carboxyhemoglobinemia The presence of carboxyhemoglobin in the blood.

carboy A large glass bottle, usually protected by a crate.

carcinogen A substance or physical agent that is capable of causing or producing cancer.

carcinogenesis Any process that produces cancer.

carcinogenic Cancer producing.

carcinoma A malignant new growth made up of epithelial cells tending to infiltrate the surrounding tissues and give rise to matastases.

cardiomyopathy A diagnostic term designating primary myocardial disease.

cardiopulmonary Relating to the heart and lungs.

cardiopulmonary resuscitation The reestablishment of heart and lung function by mechanically oxygenating and circulating the blood in cases of cardiac arrest.

cardiotoxin Any substance that produces low blood pressure and heart failure.

cardiovascular Pertaining to the heart and blood vessels.

cardiovascular disease Diseases of the heart and/or blood vessels.

carelessness Unwary, neglectful, heedless, or negligent behavior with general indifference to the safety aspects of the task or situation.

cargo manifest Shipping papers containing a listing of all the contents being carried by a transport vehicle or vessel.

carpal tunnel A passage in the wrist through which the median nerve and many tendons pass from the forearm to the hand.

carpal tunnel syndrome [CTS] An affliction resulting from the compression of the median nerve in the carpal tunnel. The carpal tunnel is a passage in the wrist through which the median nerve as well as many tendons pass from the forearm to the hand.

carrier An individual who harbors in the body the specific infectious organisms of a disease without having manifest symptoms, and thus serves as a carrier or distributor of the infection.

carrier gas A high-purity gas, primarily helium or nitrogen, that is used in gas chromatography or other processes to sweep another gas or vapor into or through a system.

cartilage Specialized, elastic connective tissue that is attached to articular bone surfaces.

cartridge (Respirator) A small container with a filter, sorbent, catalyst, or a combination of these, used for removing specific contaminants from the air passed through the container.

CAS Chemical Abstract Service.

cascade impactor An impaction-type device used for collecting airborne particulate samples on a series of impingement stages to effect a separation of the particulates by size.

case (Epidemiology) A person in the population or study group who is identified as having the disease or condition under investigation.

case control study (Epidemiology) An epidemiology study which starts with the identification of individuals with a disease or adverse health effect of interest and a suitable control group without the disease.

case fatality rate (Epidemiology) The proportion of cases of a specified condition which are fatal within a specified time.

case history (Epidemiology) The collected data concerning an individual, family, or environment, including medical history, environmental conditions, and other information useful in analyzing and diagnosing the data for the study or research being carried out.

CAS number Chemical Abstracts Service Registry number, which is a unique identification number that is assigned to each chemical. The Chemical Abstracts Service is a division of the American Chemical Society.

catabolism Any destructive process by which complex substances are converted by living cells into more simple compounds.

catalyst A substance that changes the speed of a chemical reaction, but undergoes no permanent change itself.

catalytic sensor (Instrument) A sensor with heated active and reference elements (e.g., platinum wires). The heat of combustion of the contaminant on the active element produces an imbalance in a bridge circuit such that the amount of imbalance is proportional to the concentration of the contaminant in the sampled air. This type of sensor can detect and measure the concentration of combustible gases or vapors well below their lower flammable/combustible limit.

cataract A clouding of the crystalline lens of the eye so that the passage of light is obstructed, thereby obscuring vision.

catarrh The inflammation of mucous membranes, especially the air passages of the head and throat, with free discharge.

catastrophe (OSHA) The hospitalization of five or more employees as a result of an employment accident or illness.

catastrophic release (OSHA) A major uncontrolled emission, fire, or explosion involving one or more highly hazardous chemicals that presents serious danger to employees in the workplace.

catenary The curve assumed by a flexible line under its own weight between two fixed points.

cathode Negative electrode. The electrode to which positive ions are attracted.

cathode ray tube [CRT] A vacuum tube in which an electron beam is directed at a phosphor-coated screen. The component of a video display terminal that generates the display.

cation A positively charged ion.

causal Pertaining to or denoting a cause.

causal factors A combination of simultaneous or sequential circumstances that contribute directly or indirectly to an accident, occupational disease, or other effect.

causality The relating of causes to the effects produced.

cause specific rate (Epidemiology) A rate that specifies events, such as deaths, according to their causes.

caustic An inorganic or organic chemical that is usually corrosive to human tissue, has a pH of more than 7, neutralizes acids to form salts, dissociates in water to yield hydroxide ions, and is often referred to as a base or caustic.

cavitation A condition that may occur in liquid handling equipment (e.g., pumps) in which the system pressure decreases in the suction line, pump inlet lowers fluid pressure, and vaporization occurs. Vibration, noise, and/or physical damage to equipment can result.

CAV system (Ventilation) Constant air volume system.

CBC Complete blood count.

cc Closed cup.

cc Cubic centimeter.

cd Candela.

CDC Center for Disease Control.

cd/m² Candela per square meter.

Ce Coefficient of entry.

CEEL Community Emergency Exposure Limit.

CEF Cellulose ester filter.

CEFIC European Council of Chemical Manufacturers Federation.

CEI Certified Environmental Inspector.

ceiling exposure limit value An OSHA standard that sets the maximum concentration of a contaminant a worker may be exposed to. Also referred to as a ceiling limit. The ACGIH has established ceiling limits for some substances as part of its threshold limit value table (TLV-C).

ceiling limit [C] The maximum concentration of a toxic substance that should not be exceeded in the working environment where an exposure to it may occur. The concentration of an airborne substance that is not to be exceeded at any time during the work day.

ceiling plenum A space between a suspended ceiling and the floor above it. Typically accommodates the mechanical and electrical services for a building, and often serves as a part of the ventilation system, such as the return air plenum.

ceiling value *See* ceiling limit.

cell The fundamental unit of structure and function in organisms.

cell life (Instrument) The time period over which an instrument detector can reasonably be expected to meet the performance specifications for the device.

cell membrane The thin outer layer of a cell, consisting mostly of lipids and proteins.

Celsius [C] A temperature scale that has replaced the previous designated centigrade scale of temperature.

CEM Continuous emission monitoring.

cementitious material (Asbestos) Friable materials that are densely packed and nonfibrous.

censored data Monitoring results that are nonquantitated because they are less than the limit of detection.

Centers for Disease Control and Prevention [CDC] A federal health agency and branch of the Department of Health and Human Services that provides health and safety guidance, statistical data on health issues, and training in the control of infectious diseases.

centi [C] Prefix indicating 1 E $^{-2}$.

centigrade Refers to the centigrade temperature scale. Presently, the term *Celsius* is preferred.

centimeter-gram-second system [cgs system] A coherent system of units for mechanics, electricity, and magnetism, in which the basic units of length, mass, and time are the centimeter, gram, and second.

centipoise [cp] One one-hundredth of a poise. The poise is the metric system unit of viscosity and has the dimensions of dyne-second per square centimeter.

centistoke [cSt] One one-hundredth of a stoke, the kinematic unit of viscosity. It is equal to the viscosity in poise divided by the density of the fluid in grams per cubic centimeter, both measured at the same temperature.

central nervous system [CNS] The portion of the vertebrate nervous system consisting of the brain and the spinal cord.

central processing unit [CPU] That part of a computer system that contains the control, memory, and arithmetic units.

centrifugal collector *See* cyclone as an example of this type particulate collector.

centrifuge Apparatus consisting of a compartment spun about a central axis at high speed to separate contained materials (solids or liquids) of different densities.

CEPP (EPA) Chemical Emergency Preparedness Program.

CEQ Council on Environmental Quality.

CERCLA The Comprehensive Environmental Response, Compensation, and Liability Act. Also referred as the Superfund Act.

cerebrospinal Of or pertaining to the brain and spinal cord.

certified gas-free The status of a tank, compartment, or container on a vessel that has been tested, using an approved testing instrument, and proved to be sufficiently free, at the time of the test, of toxic or explosive gases or vapors. This testing is done for a specified purpose, such as hot work, by an authorized person, and a certificate to this effect is issued.

Certified Health Physicist [CHP] An individual who has been certified in this discipline by the American Board of Health Physics.

Certified Industrial Hygienist [CIH] An industrial hygienist who has met the education, experience, and examination requirements of the American Board of Industrial Hygiene and possesses current ABIH certification as an industrial hygienist (i.e., has been certified as competent in one or more aspects of this discipline by the American Board of Industrial Hygiene).

Certified Safety Professional [CSP] A safety person who has been certified in one or more aspects of this discipline by the Board of Certified Safety Professionals.

cerumen Ear wax.

CES Certified Environmental Specialist.

CET Certified Environmental Trainer.

CET Effective temperature corrected for radiation.

cf. Compare.

CFC Chlorofluorocarbons.

cfh Cubic feet per hour.

cfm Cubic feet per minute.

cfm/sq ft Cubic feet per minute per square foot.

CFR U.S. Code of Federal Regulations.

CFU Colony forming units.

CFU/m³ colony forming units per cubic meter.

CGA Compressed Gas Association.

CGI Combustible gas indicator.

cgs system Centimeter-gram-second system.

chain of custody form A form used for tracking samples from the time the samples are obtained, through their transportation, receipt at the laboratory, and analysis.

chain reaction (Nuclear Industry) A self-sustaining reaction that stimulates its own repetition wherein a fissionable nucleus absorbs a neutron and fissions, releasing additional neutrons that are absorbed by other fissionable nuclei, releasing still more neutrons.

chalcosis The presence of copper deposits in tissues.

chalicosis A pnemoconiosis resulting from the inhalation of fine particles of stones. Also referred to as flint disease.

chalkitis Inflammation of the eyes as a result of rubbing the eyes with the hands after the hands have contacted brass. Also referred to as brassy eye.

challenge agent (Respiratory Protection) The aerosol, gas, or vapor used to determine the effectiveness of a respirator fit.

charcoal tube A glass tube of specified dimensions and assembly, containing 100 mg of 20/40 mesh activated coconut shell charcoal in a front section and 50 mg in a backup section. Larger tubes are available.

charged particle An elementary particle carrying a positive or negative electric charge.

Charles' law A physical law that states that, at constant volume, the pressure of a confined gas is proportional to its absolute temperature.

CHD   Coronary heart disease.

check valve   A self-closing device which is designed to permit the flow of fluids in one direction and to close if there is a reversal of flow.

chelate   A coordination compound in which a central metal ion is attached by coordination links to two or more nonmetal atoms in the same molecule.

chelating agent   A compound which will inactivate a metallic ion with the formation of an inner ring structure in the molecule, with the metal ion becoming a member of the ring. Some are used to remove metals, such as lead, from the body.

chelation   The formation of a chemical ring structure containing a metal ion that is complexed by linkage to two or more nonmetal atoms in the same molecule.

chelation therapy   The treatment of lead poisoning with chemical agents which will bind the lead in the blood and enable it to be excreted in the urine.

chemical   Any element, chemical compound, or mixture of elements and/or compounds.

chemical absorption detector   *See* detector tube.

Chemical Abstracts Service   [CAS]   A division of the American Chemical Society that provides a systematic computerized source of chemical information, such as the Chemical Compound Registry, and which assign unique numbers to each chemical substance.

chemical agent   A hazardous substance, chemical compound, or a mixture of these.

chemical asphyxiant   A chemical material which has the ability to render the body incapable of utilizing an adequate oxygen supply even though there is a normal amount of oxygen in the inspired air.

chemical burn   A burn, generally the same as that caused by heat.

chemical cartridge   A device containing an adsorbent and/or absorbent material for use in respirators for removing gases and/or vapors from inspired air.

chemical cartridge respirator   A respiratory protective device that is equipped with a cartridge(s) for removal of low concentrations of specific vapors or gases.

chemical composition   The identity and percent by weight of each element in a compound.

chemical formula   The number and symbol of each of the atoms making up a compound.

chemical fume hood   An exhaust-controlled enclosure with one open side that is provided with a sliding closure (sash) and internal baffles to distribute airflow across the open face.

chemical manufacturer   A person/business who imports, produces, or manufactures a chemical substance for use or distribution.

Chemical Manufacturer's Association   [CMA]   An association of chemical product manufacturers that disseminates information on the safe handling, transportation, and use of chemicals. In addition, it develops labeling guidelines and provides medical advice on the prevention and treatment of chemical injuries.

chemical oxygen demand   [COD]   A measure of the oxygen required to oxidize all oxidizable (organic and inorganic) components present in water.

chemical pneumonitis   Pneumonitis or inflammation of the lung parenchyma as a result of the aspiration of a hydrocarbon solvent which spreads rapidly as a film over the lung's surfaces. The inhalation of beryllium or cadmium fumes or dust can cause an acute pneumonitis.

chemical reaction A change in the arrangement of atoms and molecules to yield substances of different composition and properties.

chemical sensitivity The sensitivity that some people experience when exposed to certain chemicals, resulting in health problems, such as dermatitis, nasal congestion, and others. For example, many people are sensitive to nickel.

chemical structure The molecular structure of a compound.

Chemical Transportation Emergency Center [CHEMTREC] A section of the Chemical Manufacturer's Association that provides emergency response information upon request to control an emergency.

chemical waste The waste generated by chemical, petrochemical, plastic, pharmaceutical, biochemical, or microbiological manufacturing processes.

chemiluminescence The emission of absorbed energy as light, due to a chemical reaction of the components of the system. This principle is employed in some instruments for determining the airborne concentration of some substances (e.g., ozone).

chemiluminescent detector A detector that is designed to detect light produced in chemical reactions, such as that between ozone and ethylene or nitric oxide. This phenomenon is employed in determining ambient levels of ozone and oxides of nitrogen.

chemisorb To take up and hold, usually irreversibly, by chemical forces.

chemisorption The collection of a contaminant by a sorbent as a result of the formation of bonds between a material with high surface energy and a gas, vapor, or liquid in contact with it.

chemosterilant A chemical that controls pests by preventing reproduction.

chemotherapy The treatment of disease using chemical agents.

CHEMTREC Chemical Transportation Emergency Center.

Cheyne-Stokes respiration A form of respiration in which the individual appears to have stopped breathing for up to 40–50 seconds, then breathing starts again with increasing intensity, then stops as before, and then repeats the previous breathing rhythm.

CHF Congestive heart failure.

chilblain A recurrent localized itching, swelling, and painful erythema of the fingers, toes, or ears as a result of a mild frostbite to the tissues.

chill temperature The perceived temperature (as opposed to the actual temperature) due to the cooling power of wind on exposed flesh. Also referred to as the Wind Chill Factor.

China syndrome The hypothetical result of a nuclear power reactor core meltdown in which the molten fuel melts through the reactor pressure vessel and the bottom of the containment building and into the earth all the way to China.

chlamydia A family of organisms that cause a wide variety of diseases in man, such as psittacosis.

chloracne An acneiform (resembles acne) eruption resulting from skin contact with halogenated aromatic compounds.

chlorofluorocarbon A compound containing carbon and one or more halogens, usually fluorine, chlorine, or bromine.

chlorophyll   Any of a group of related green pigments found in photosynthetic organisms.

CHMM   Certified Hazardous Material Manager.

cholinesterase   An enzyme that catalyzes the hydrolysis of acetylcholine to choline and an anion. Essential for proper nerve function.

cholinesterase inhibition   The loss or decrease of enzymatic activity of cholinesterase caused by binding of the enzyme with another chemical.

CHP   Certified Health Physicist.

CHRIS   Chemical Hazard Response Information System.

chromatin   The more readily stainable portion of a cell nucleus.

chromatogram   For the differentiating type detector, which is the most common type, the chromatogram is a graphical presentation corresponding to the components present in the sample introduced into the instrument. The elapsed time from sample injection to each peak is a means to identify the components in the sample. The area under each peak is proportional to the total mass of that component in the sample. The chromatogram for the integrating type detector is a series of plateaus, with each plateau proportional to the total mass of the component in the eluted zone.

chromatographic detector   One of two types of detectors. The integrating type detector gives a response proportional to the total mass of component in the eluted zone, while the differentiating type gives a response proportional to the concentration or mass flow rate of the eluted component.

chromatography   An analytical method used to separate the components of gas, liquid, and/or vapor mixtures based on selective adsorption by which the components of complex mixtures can be identified.

chromosome   Chromosomes are constituents of the nucleus of all cells and contain a thread of DNA, which transmits genetic information. They control the reproduction of cells that are produced from the original cell.

chronic   Long lasting, persistent, prolonged, repeated, or frequently recurring over a long period.

chronic disease   A disease that is slow in its progress and of long duration in developing.

chronic effect   An effect that is the result of exposure to a toxic substance over a long period. The daily dose is insufficient to elicit an acute response, but it may have a cumulative effect over a period of time. Often, the rate of absorption of the toxic agent exceeds the rate of elimination, thereby resulting in a buildup of the substance in the body.

chronic exposure   Exposure for lengthy periods ranging from months to years. The exposure is usually frequent or continuous throughout the work day.

chronic obstructive pulmonary disease   A chronic obstruction of bronchial airflow resulting in decreased pulmonary ventilation.

chronic toxicity   The toxic effect of a chemical substance or physical agent that results from repeated or persistent exposure to the substance or agent.

chrysotile asbestos   Asbestiform mineral in the serpentine group that has been used as an insulation material in buildings. Referred to as white asbestos, it is the type that has been the most widely used in the U.S.

Ci   Curie.

CIH   Certified Industrial Hygienist.

CIIT   Chemical Industry Institute of Toxicology.

cilia   Plural of a cilium, which is a minute hairlike process attached to the free surface of a cell. The cilia on the surfaces of respiratory passages are tiny hairlike appendages that aid in removing particulates that collect on their moist surfaces.

circadian rhythm   The rhythmic repetitions of certain phenomena in living organisms that occur at about the same time in each 24-hour day.

circumoral paresthesia   A burning sensation around or near the mouth.

cirrhosis   Liver disease characterized pathologically by loss of the normal microscopic architecture of this organ with fibrosis and nodular regeneration.

citation   A notice issued by a regulatory agency alleging specific conditions which are in violation of a regulatory standard.

CIUB accident   An accident in which the individual is caught in, under, or between an object or objects.

claims-made policy   Insurance policy providing coverage of an occurrence if a claim is made during the term of the policy.

Class I biological safety cabinet   An open-fronted, negative pressure, ventilated cabinet with a minimum inward face velocity at the work opening of at least 75 feet per minute with the exhaust air filtered through a HEPA filter.

Class II biological safety cabinet   (laminar flow)   An open-fronted, ventilated cabinet with an average inward face velocity at the work opening of at least 75 feet per minute, providing HEPA filtered recirculated airflow in the cabinet work-space and exhaust air passed through a HEPA filter.

Class III biological safety cabinet   A totally enclosed cabinet of gas-tight construction, such as a glove box. The exhaust fan for this cabinet is a dedicated unit with exhaust air discharged directly to the outdoors. Air entering the cabinet is passed through a HEPA filter. Operations are conducted in the enclosure using glove ports. In use, the cabinet is maintained at 0.5 inches water gauge negative pressure.

Class 100 Clean Room   An area/room in which the particle count in the air does not exceed 100 particles per cubic foot in the size range of 0.5 micrometers and larger.

Class 10,000 Clean Room   An area/room in which the particle count in the air does not exceed 10,000 particles per cubic foot larger than 0.5 micrometers or 65 particles per cubic foot larger than 5 micrometers in size.

Class 100,000 Clean Room   An area/room in which the particle count in the air does not exceed 100,000 particles per cubic foot larger than 0.5 micrometers or 700 particles per cubic foot larger than 5 micrometers.

Class I Laser   A laser also referred to as an exempt laser. Under normal conditions, this laser does not emit a hazardous level of optical radiation.

Class II Laser   A low power laser which may cause retinal injury if viewed for long periods of time.

Class III-A Laser   A visible laser that can cause injury to the eyes. Class III laser devices are classed as medium-power laser devices.

Class III-B Laser A laser that can cause injury to the eye as a result of viewing the direct or reflected beam. Class III laser devices are classed as medium-power laser devices.

Class IV Laser These high-power laser systems require extensive exposure controls for preventing eye and skin exposure to both the direct and reflected laser beam.

Class V Laser Includes any Class II, III, or IV laser device which, by virtue of appropriate design or engineering controls, cannot directly irradiate the eye at levels in excess of established exposure limits.

classes of carcinogens *See* A1, A2, A3, A4, and A5 carcinogen listed under A.

Classes of fires *See* fires.

clastogenic Substance that damages chromosomes.

clean air (OSHA) Air of such purity that it will not cause harm or discomfort to an individual if it is inhaled for extended periods of time.

clean air (Sampling) Air that is free of any substance that will adversely affect the operation of or cause a response of an instrument.

Clean Air Act [CAA] An EPA regulation intended to protect the quality of the nation's air from further deterioration and to improve its quality in certain areas.

clean area (Asbestos Abatement) A controlled environment that is maintained and monitored to assure a low probability of asbestos contamination in that space.

clean facial policy *See* facial hair policy.

clean room A specially constructed area or space that is controlled for airborne aerosols, temperature, humidity, air flow, and air pressure. *See* Class 100/10,000/100,000 Clean Room.

clean-up operation (OSHA) An operation in which hazardous substances are removed, contained, incinerated, neutralized, stabilized, cleared-up, or in any other manner processed or handled with the ultimate goal of making the site safe for people or the environment.

clearance (Physiology) Complete removal of a substance from a specific volume of blood per unit of time.

clearance sampling (Asbestos Contamination Control) A sampling procedure carried out at the end of an asbestos abatement activity to determine whether the asbestos abatement has been effective and the fiber concentration is acceptable. Typically, the acceptable concentration is the background level, or that which has been specified in the abatement contract.

Cleveland open-cup test [COC] A method for determining the flash point of a material.

climatic Relating to the meteorological conditions, including temperature, precipitation, and wind, that characteristically prevail in a particular region.

clo A unit of clothing insulation necessary to keep a sitting man comfortable in a ventilated room at 70°F and 60 percent relative humidity.

closed circuit SCBA Self-contained respiratory protective device in which the breathing air is recirculated and rebreathed after carbon dioxide has been removed to maintain the quality of the breathing air.

closed cup-test [CC] A method for determining the flash point of a flammable liquid.

closed face sampling Sampling with filter media using a two or three piece cassette with the inlet section in place and the sealing plugs removed.

closed installation (Laser) A location where lasers are used which will be closed to unprotected personnel during laser operation.

cluster (Epidemiology) An increased incidence or suspected excess occurrence of a disease in time, location, area, occupation, etc.

cluster (Fibers) A structure with fibers in a random arrangement such that all fibers are intermixed and no single fiber is isolated from the group, and with the groupings having more than two intersections.

cm Centimeter.

CMA Chemical Manufacturer's Association.

CMOS Complimentary metal oxide semiconductor sensor.

CNC Condensation nuclei counter.

CNG Compressed natural gas.

CNS Central nervous system.

CNS effect Effects which occur to the central nervous system, including drowsiness, dizziness, loss of coherence and reasoning, and others.

coagulation The formation of a clot or gelatinous mass.

coalesce To join together, such as in the union of two or more droplets of a liquid to form a larger droplet.

coalition for occupational safety and health Grassroots coalition of occupational health activists, local trade unions, and environmental groups.

coal miner's pneumoconiosis A pneumoconiosis resulting from the deposition of coal dust in the lungs. Characterized by emphysema. Also referred to as black lung and coal worker's pneumoconiosis.

coal-tar pitch volatiles Term used in reference to the emissions from coke ovens and for which OSHA has established a PEL.

co-carcinogen An agent which increases the effect of a carcinogen by direct concurrent local effect on the tissue.

cochlea The essential organ of hearing shaped like a spiral wound tube resembling a snail's shell. The auditory part of the inner ear.

COC test *See* Cleveland open-cup test.

COD Chemical oxygen demand.

Code of Federal Regulations [CFR] The rules promulgated under U.S. law and which are published in the Federal Register. A codification of the general and permanent rules published in the Federal Register by the executive department and agencies of the federal government that is divided into 50 titles representing areas subject to federal regulation.

coefficient of entry [Ce] The actual rate of flow caused by a given hood static pressure compared to the theoretical flow which would result if the static pressure could be converted to velocity pressure with 100 percent efficiency.

coefficient of haze [COH]   A measure of the haziness of the atmosphere expressed in COH units, which are 100 times the optical density.

coefficient of variation [CV]   Statistical parameter equal to the standard deviation of the sample data divided by the mean of the data. Often expressed as the percent coefficient of variation. Another term for it is the relative standard deviation.

cogeneration   The simultaneous production of electricity and steam from the same energy source.

COH   Coefficient of haze.

COHb   Carboxyhemoglobin.

coherent   Composed of waves having a continuous relationship among phases or waves in phase in both time and space, such as a light beam.

cohesion   The molecular forces of attraction between particles of like composition.

cohort   A group of individuals selected for scientific study in toxicology or epidemiology.

cohort effect (Epidemiology)   The systematic differences among two cohorts with respect to the distribution of some variable such as age, weight, etc.

cohort study (Epidemiology)   A method of epidemiologic study in which subsets of a defined population can be identified who are, have been, or in time may be exposed or not exposed to a factor or factors which may influence the probability of the occurrence of a given disease or other outcome.

cold work   Mechanical or other type work of a nonsparking nature that presents no fire/explosion risk.

colic   A severe cramping in the abdomen.

coliform index   A rating of the purity of water based on a count of fecal bacteria present in it.

coliform organism   Microorganisms found in the intestinal tracts of humans and animals. Their presence in water and foods indicates potentially dangerous bacterial contamination.

collagen   A main supportive protein of skin, tendon, bone, cartilage, and connective tissue.

collection efficiency   A measure of sampler performance as determined from the ratio of the material collected to the amount present in the sampled air. Typically expressed as a percentage.

collimated beam   A beam of light or electromagnetic radiation with parallel waves. Also referred to as collimated light.

collimator   A device for limiting the cross-sectional area or confining the beam within a solid angle, such as with X-ray machines and radiation sources.

colony-forming unit   Biological organisms present in the air that give rise to the formation of colonies when mixed with a nutrient, placed in a petri dish, and incubated for an appropriate period. A method for determining the number of viable organisms per unit quantity of air.

colophony   Rosin, such as that used in rosin core solder.

colorimetry   An analytical method in which color is developed in a reaction between the sorbent and a contaminant, with the resulting color intensity measured photometrically for determining contaminant concentration. *See* photometry.

coma   A state of unconsciousness from which the person cannot be aroused by physical stimulation.

combustible   Capable of being ignited with resultant burning or explosion.

combustible dust   A dust that is capable of undergoing combustion or of burning when subjected to a source of ignition.

combustible gas indicator   [CGI]   An instrument for determining the presence and concentration of a combustible/flammable hydrocarbon vapor/gas-air mixture relative to the lower explosive limit of the substance. Essentially all combustible/flammable vapors or gases can be detected with this type device, but their concentration cannot be determined accurately unless the instrument has been calibrated for the specific substance/mixture. It is essential that adequate oxygen be present (i.e., above about 12 percent) for the proper operation of this type detector.

combustible liquid   A liquid having a flash point at or above 100°F. There are three classes of combustible liquids: Class II with flash points above 100°F and below 140°F; Class IIIA with flash points at or above 140°F and below 200°F; and Class IIIB includes those with flash points at or above 200°F.

combustion   A chemical process that involves oxidation sufficient to produce light or heat.

combustion air (Acoustics)   The air used for the combustion of fuel in a furnace. Typically furnace burners were provided with primary and secondary air inlets which contributed to furnace noise. Present design provides for forced air to the burners, which reduces noise emissions for the combustion of the fuel.

combustion gases   The mixture of gases produced by the combustion process. Typical emissions include carbon dioxide, carbon monoxide, oxides of nitrogen, and others.

combustion products   Materials produced or generated during the burning or oxidation of a substance.

comfort ventilation   Airflow intended to remove heat, odors, smoke, etc. from an inside location and provide a comfortable environment for occupants.

comfort zone   The range of effective temperatures, as identified by ASHRAE, over which the majority (50 percent or more) of adults feel comfortable. ASHRAE has identified combinations of dry- and wet-bulb temperatures and air movement for summer and winter conditions that provide comfort for room occupants.

commissioning   The initial acceptance process in which the performance of equipment/ system is evaluated, verified, and documented to assure its proper operation in accordance with codes, standards, design specifications, etc.

common law   The system of laws that originated in England based on court decisions, the doctrines implicit in those decisions, and customs and usages, rather than on codified written law.

common name   The designation of a material by other than its chemical name, such as by a code name or generic name.

communicable   As applied to disease, it is one that results from the spread or transmission of an infectious agent. The causative agent of the disease can be transmitted from an infected individual to another. Some diseases of animals are transmissible to man and are thus considered communicable diseases.

communicable disease   An illness due to an infectious agent or its toxic products which are transmitted directly or indirectly to a well person from an infected person or animal, or through the agency of an intermediate animal host vector or an inanimate environment.

communicable period   The time during which an infectious agent may be transferred directly or indirectly from an infected person to another person, from an infected animal to a person, or from an infected person to an animal.

Community Awareness and Emergency Response Program   [CAER]   A program developed by the CMA to provide help to plant managers to take the initiative to cooperate with local communities in the development of integrated hazardous materials response plans.

Community Right to Know Act   A federal act requiring states to establish emergency response commissions, emergency planning committees and districts, and comprehensive emergency planning programs to respond to a hazardous chemical release, and requiring industries to report the presence of or release of hazardous chemicals.

compensable injury   An occupational injury or occupational disease resulting in sufficient disability to require the payment of compensation as prescribed by law.

competent person (OSHA)   One who knows the hazards existing or likely to exist, knows how to control or eliminate those hazards, and has been given the authority to correct the hazards promptly and does so.

complainant   An individual who registers a complaint to OSHA, etc. concerning working conditions or any other unsatisfactory situation.

complete blood count   [CBC]   A measure of the hemoglobin concentration and the numbers of red blood cells, white blood cells, and platelets in one cubic millimeter of blood. In addition, the proportion of various white blood cells is determined, and the appearance of red and white cells is noted.

complete combustion   Fuel burning in which all the combustible is burned.

compliance monitoring   A strategy or technique of monitoring or determining compliance with a government standard. One compliance monitoring method is to identify the maximally exposed worker. If that exposure is less than the standard, then all worker exposures are assumed to be below the exposure limit.

compliance officer (OSHA)   Federal compliance safety and health officer. *See* Compliance Safety and Health Officer.

compliance program   Typically, a written program which identifies the methods and procedures that will be implemented to comply with a regulatory standard.

Compliance Safety and Health Officer (OSHA)   [CSHO]   Individual empowered to carry out the enforcement and monitoring aspects of the Occupational Safety and Health Act under the direct supervision of an area director.

compliance strategy   The method an employer will develop and implement to achieve and maintain compliance with a regulation. It may include engineering and administrative controls, adherence with established procedures and work practices, the use of personal protective equipment, as well as training of personnel regarding hazards and making available to personnel hazard information (i.e., signs, material safety data sheets, hazard communication training, etc.).

complimentary metal oxide semiconductor   [CMOS]   A type of detector used in the detection of gases or vapors.

composite material   A stiff, strong, and lightweight material consisting of a matrix, such as epoxy, reinforced by strong, stiff filaments, such as carbon.

compound   A substance composed of atoms or ions of two or more elements in chemical combination. The constituents are bound by bonds or valence forces.

Comprehensive Environmental Response, Compensation, and Liability Act   [CERCLA] An EPA act for remedying the release of hazardous substances and for addressing the cleanup of sites in which hazardous wastes have been disposed.

Compressed Air and Gas Association   An association of compressed air and gas manufacturers and suppliers.

compressed gas   Any material or mixture having in the container an absolute pressure exceeding 40 pounds per square inch at 70°F or, regardless of pressure at 70°F (21.1°C), an absolute pressure exceeding 104 pounds per square inch at 130°F, regardless of the pressure at 70°F; or any liquid material having a vapor pressure exceeding 40 pounds per square inch absolute at 100°F as determined by ASTM Test D-323.

Compressed Gas Association   [CGA]   Association of gas producers, suppliers, equipment manufacturers, and related industries that develop safety standards, make recommendations to improve methods of handling, transporting, storing of compressed gases, and advise regulatory agencies concerned with the safe handling of compressed gases.

compressed gas in solution   A nonliquefied compressed gas which is dissolved in a solvent.

Compton scattering   One of the processes by which radiation loses energy to matter. It involves the transfer of part of the energy of an electron to the matter it is passing through, as well as the deflection of the electron from its original path. An electron interacts with an orbital electron of an atom to produce a recoil electron and a scattered photon.

computer vision syndrome   Computer-related vision and eye problems, such as eyestrain and blurred vision, that result from extended work at a computer screen.

CONCAWE   Conservation of Clean Air and Water in Europe.

concentration   The quantity of a substance contained in a unit quantity of sample. Expressed as parts per million (ppm), milligrams per cubic meter (mg/m^3), fibers per cubic centimeter (f/cc), etc.

concession   A tract of land granted by a government to an individual or a company for exploration and exploitation in recovering minerals.

concomitant   Occurring at the same time.

condensation   The change of state of a substance from the vapor to the liquid or solid form.

condensation nuclei   Small particles on which water vapor condenses.

condenser (Microscopy)   Any lens or system that is designed to gather light rays and cause them to converge at a focus point.

conditioned air   Air that has been heated, cooled, humidified, or dehumidified to maintain an interior space within the comfort zone. Sometimes referred to as tempered air.

conditioned space (Ventilation)   That part of a building which is heated or cooled, or both, for the comfort of occupants.

conductance   A measure of a material's ability to conduct electrons.

conduction (Heat Stress)   The transfer of heat by physical contact between bodies.

conductive hearing loss   A type of hearing loss that is not caused by noise, but is due to any disorder that prevents sound from reaching the inner ear. A hearing loss that is due to poor transmission of sound from the outer ear to the cochlea.

conductivity detector   A device that measures electroconductivity using a method based on the absorption of a gas by an aqueous solution with the formation of electrolytes, thereby producing a change in the electroconductivity of the solution, which is measured and equated to gas concentration.

confidence (Analytical Instrument Response)   The degree to which a measurement is believed to be true.

confidence interval   A range or interval that has a specified probability of including the true value of a parameter of a distribution.

confidence limits   Confidence limits are mathematically determined intervals defined as upper and lower limits, in which one is confident (e.g., 90 percent, 95 percent, etc.) that the true value is greater than, less than, or between.

confidentiality   The state of being done or communicated in confidence. The right of a person to withhold information from another.

confined space   A space which has limited openings for access/egress and unfavorable natural ventilation, is not intended for continuous occupancy, and could contain or produce a hazardous atmosphere. A confined space may include, but not be limited to, a tank, pit, boiler, or sewer utility vault.

confined space entry   The entry of personnel (one or more) into a confined space.

confined space-permit required (OSHA)   [PRCS]   A confined space that contains or has the potential to contain a hazardous atmosphere or a material with the potential to engulf an entrant, is configured such that an entrant could be trapped or asphyxiated, or contains any other recognized health or safety hazard.

conflict of interest   The compromising of a person's objectivity when that person has a vested interest in the outcome.

confounder   A factor which may not have been taken into account in a study and thus results in confusion.

confounding   A situation in which the effect of an exposure or risk is distorted due to an association of the exposure with other factors that influence the outcome.

confusion   An early stage of psychosis involving mental and emotional disturbances.

congenital   Existing at birth, but not due to heredity.

congenital malformation   An abnormality that is present at birth.

congruent   Coinciding exactly when superimposed.

conjugation (Toxicology)   The combination of a toxic product with some substance in the body to form a detoxified product, which is then eliminated.

conjunctiva   The mucous membrane that lines the eyelids and covers the exposed surface of the eyeball.

conjunctivitis   Inflammation of the conjunctiva, which is the delicate membrane that lines the eyelids and covers the exposed surface of the eyeball.

consensus standard   A standard developed through a consensus process of agreement among representatives of interested or affected industries, organizations, or individual members of a nationally recognized standards-producing organization. The process includes peer review and public comment, resulting in a standard that reflects agreement among participants with diverse interests and expertise.

constant air volume system   A ventilation system that provides a constant supply of air of varying temperature to meet the heating and cooling needs of an area.

construction work (OSHA)   Work for construction, alteration, or repair, including painting and decorating.

contact   A person or animal that has been in such association with an infected person, animal, or a contaminated environment as to have had opportunity to acquire the infection.

contact dermatitis   A dermatitis that is caused by contact with a primary irritant.

contact, primary   Persons in such direct contact or association with a communicable disease case as to have had the opportunity of acquiring the infection.

contact, secondary   Person who is in contact with or who has associated with a primary contact.

contagion   Disease transmission by direct or indirect contact. Literally, the transmission of infection by contact.

contagious   Capable of being transmitted from one individual to another by direct or indirect contact; spreading or tending to spread from one to another.

contagious disease   A disease that is highly infectious and is transmitted by contact.

containment   Isolation of a work area from the rest of the building/structure to prevent the spread of contaminants to other areas.

containment system   The system, including the structure, ventilation method, entry/egress routes, contaminant collection equipment, etc., that will be utilized to prevent the spread of contamination from a work site into the surroundings.

contaminant   A harmful, irritating, or nuisance-producing material that is foreign to the normal environment. A substance whose presence in the air is harmful, hazardous, or deleterious.

contamination   The presence of a potentially harmful substance in the air, on a surface, on the skin, or in a material.

contingency plan   A document indicating an organized, planned, and coordinated course of action to be followed in case of an emergency or situation that could affect the health or welfare of the community or the environment.

continuous air monitor   [CAM]   An instrument which is typically located in a potentially contaminated location to detect a specific contaminant, such as a flammable or toxic gas or vapor, and which will alarm if a preset concentration is exceeded. It can be a passive type sampler or an active type sampler.

continuous exposure   Exposure to a health hazard throughout the workday.

continuous monitoring system   The taking and recording of sample results continuously. Such methods are typically employed in sampling systems for air and water.

continuous noise   Variations in noise level involving maxima at intervals of 1 second or less. Also defined as broadband noise of approximately constant level and spectrum, to which an employee is exposed for a period of 8 hours per day, 40 hours per week.

continuous noise (General)   Broadband noise of approximately constant level and spectrum to which an employee is exposed for a period of 8 hours per day, 40 hours per week.

continuous spectrum (Acoustics)   A spectrum which is continuous in the frequency domain.

continuous wave (Laser)   [CW]   A laser system which provides a constant, steady-state delivery of laser power.

contour   Line connecting points of equal magnitude, such as noise contours or elevation contours.

contour map   A map that has lines marked to indicate points or areas that are the same elevation above or below sea level.

contributory negligence   Negligence shown to have been caused by a claimant who contributed to an injury by not exercising ordinary care; the injury is then a case of contributory negligence.

control (Epidemiology/Toxicology/etc.)   Persons in a comparison group that differs in disease experience from the subjects of the study. The nature, number, and reproducibility of the controls (unexposed or unaffected) to determine the accuracy and significance of the conclusions from the experimental (exposed) cohort results. A most important factor in any study of humans, animals, or biological organisms.

control (Industrial Hygiene)   Measures, including engineering and administrative means as well as the use of personal protective equipment, that are implemented to reduce, minimize, or otherwise control exposure to a health hazard.

control efficiency   The ratio of the amount of pollutant removed from a source of release/emission by a control device to the total amount of pollutant before control, and expressed as a percentage.

control group (Toxicology)   A group of test animals in a study which are not dosed, but are otherwise treated the same as the dosed animals.

controlled area (Ionizing Radiation)   A specified area in which exposure to radiation or a radioactive material is controlled. The controlled area should be under the supervision of an individual with the knowledge and responsibility to apply appropriate radiation protection measures to minimize exposure.

control rod (Nuclear Industry)   A rod, tube, or plate that is used to control the power produced by a nuclear reactor by preventing neutrons from causing further fission.

control system   The equipment or procedure that is in place to reduce the amount of a pollutant that is released to the workroom air or the ambient environment.

control velocity   *See* capture velocity.

contusion   An injury to a body part without a break in the skin, such as a bruise.

convection   The motion in a fluid, such as air, that results from differences in density and the action of gravity.

convective heat transfer (Heat Stress)   The transfer of heat to or from the body as well as to the surrounding air movement based on the difference in temperature between the air and the body.

convergence (Laser)   The bending of light rays toward each other.

converter   A device that changes electric current from one kind (e.g., ac or dc) to another.

convulsion   An intense involuntary muscular contraction.

coolant   A heat transfer agent used to convey heat from its source.

cooling tower   A structure that is used for the removal of heat from water used for cooling in industrial operations, such as petroleum refineries, petrochemical facilities, electric power generating plants, etc.

copolymer   A long chain molecule resulting from the reaction of more than one monomer species with that of another.

coproporphyrin   A porphyrin that is formed in the blood-forming organs and found in the urine and feces.

core (Nuclear Power)   The control portion of a nuclear reactor containing the fuel elements, moderator, neutron poisons, and support structure.

core (Physiology)   Within the body as opposed to outside it.

core temperature   The temperature in the central part of the body. Rectal temperature is considered a measure of core temperature.

corium   The layer of skin deep to the epidermis, consisting of a dense bed of vascular connective tissue.

cornea   The transparent outer membrane which covers the eye.

corona   A luminous discharge due to the ionization of air surrounding a conductor that occurs when the local electric field exceeds the dielectric strength of the air.

corrective lens   An eyeglass lens that has been ground to the wearer's individual prescription to enable normal visual acuity.

correlation   The degree of association between variables. The simultaneous increase or decrease in the value of two random variables (positive correlation), or the simultaneous increase in the value of one and decrease in the value of the other (negative correlation).

correlation coefficient   A measurement of the strength of the relationship between two variables.

corrode   The gradual breaking down, wearing away, or alteration of a structure due to the action of air, moisture, or a chemical.

corrosion   A process of dissolving or wearing away gradually, especially of metals, through chemical or electrochemical action.

corrosive   Substance which chemically attacks materials with which it comes in contact.

corrosive waste   Waste having the ability to corrode standard containers or to dissolve toxic components or other waste.

corundum   Natural aluminum oxide material that may contain traces of iron, magnesium, and silica.

coryza   An acute inflammation of the nasal mucous membrane with profuse discharge.

COSH   Coalition for Occupational Safety and Health.

COSHRA   Comprehensive Occupational Safety and Health Reform Act.

cosmic radiation   High-energy particulate and electromagnetic radiations which originate outside the earth's atmosphere. Also referred to as cosmic rays.

cost benefit analysis   An economic analysis in which the cost of preventing a disease or adverse health effect is compared to the cost of a case of the disease or adverse health effect. Also the comparison of the cost of implementing a regulation or control measure to the value of the benefits received as a result of implementing the regulation or control measure.

coulomb   [C]   A quantity of electric charge equal to one ampere second.

Council for Accreditation in Occupational Hearing Conservation   [CAOHC]   Professional group involved in the development and review of an approved course leading to certification of an individual as an Occupational Hearing Conservationist.

count (Ionizing Radiation)   An indication of the ionizing events which occur in a period of time, such as counts per minute. This must be corrected based on the counter efficiency to determine disintegrations per unit time.

counter (Ionizing Radiation)   A device for counting nuclear disintegrations in order to measure the amount of radioactivity in a sample.

countermeasure   An action taken in opposition to another.

cp   Centipoise.

CP   Chemically pure.

CPC   Chemical protective clothing.

cpm   Counts per minute.

CPR   Cardiopulmonary resuscitation.

cps   Cycles per second.

CPSC   U.S. Consumer Product Safety Commission.

cradle to grave   A technique for ensuring that a hazardous waste is effectively disposed of such that it causes no environmental concerns. It involves the identification of the generator of the waste as well as its type; tracking of the waste to the disposal facility by manifest; requiring of permitting for generators, storage facilities, and disposal sites; and the enforcement of regulations to ensure compliance.

cramps   Sudden involuntary muscular contractions which cause severe pain. Painful muscle spasms in the extremities, back, or abdomen, as a result of or due in part to excessive loss of salt during sweating.

crest factor (Acoustics)   A factor that is equal to the ratio of the peak acoustic pressure to the rms value.

cristobalite   A crystalline form of silica.

criteria pollutants   Ambient air pollutants which are known to adversely affect human health.

criterion   A standard, rule, or test on which a judgment or decision can be based.

criterion sound level   A sound level of 90 decibels (A-weighted).

critical (Nuclear Power Industry)   A condition of sustained nuclear chain reaction. A nuclear reactor is critical when the rate of neutron production is equal to the rate of neutron loss.

critical barrier (Asbestos)   One or more layers of plastic sealed over at all openings into the work area sufficient to prevent airborne asbestos in the work area from releasing into adjacent areas.

critical equipment   Equipment that is likely to result in a major problem or loss if damaged, that operates improperly, or that ceases to operate for whatever cause, and is therefore considered vital to the continued effective safe operation of the system/process.

critical flow   A flow condition in which air is flowing at sonic velocity.

critical flow orifice   A device used for determining volumetric flow rate with an accuracy of between 90 and 110 percent if made to standardized dimensions.

critical illumination   Illumination necessary for microscopic examination of asbestos samples by count or for microscopic examination of other slide samples. The light source must be in focus at the front focal plane of the condenser, and the field iris must be in focus at the sample plane.

criticality (Nuclear Power)   A condition when the number of neutrons released by fission is exactly balanced by the neutrons being absorbed and escaping the reactor.

criticality (System Safety)   The seriousness of loss in terms of lives, dollars, or the effect on operations.

criticality accident   An uncontrolled chain reaction in which a very large amount of energy is liberated in a very brief period of time.

critical mass (Nuclear Industry)   The smallest amount of fissionable material that can sustain a chain reaction.

critical organ (Ionizing Radiation)   The anatomical organ which is particularly affected by the assimilation of a radioactive material; or that which is most seriously affected by external radiation.

critical organ (Toxicology)   The organ in the body which receives the greatest damage as a result of exposure to a health hazard.

critical orifice   *See* critical flow orifice.

critical pressure   The pressure required to liquefy a gas at the critical temperature.

critical speed (Vibration)   Any rotating speed which is associated with high vibration amplitude.

critical temperature   The temperature above which a gas cannot be liquefied by pressure alone.

critical velocity (Ventilation)   The velocity above which streamline flow becomes turbulent.

crocidolite asbestos   An amphibole variety of asbestos containing approximately 50 percent combined silica and nearly 40 percent combined iron (valence 2/3). This type asbestos fiber has been considered the most toxic form of asbestos by some health professionals and regulatory agencies. Often referred to as blue asbestos.

cross connection   Any physical connection or arrangement between two piping systems such that there may be a flow from a nonpotable source into the potable water supply piping.

cross-sectional study (Epidemiology)   A one-time survey of a population generally intended to describe the existing prevalence of a disease.

CRT   Cathode ray tube.

crude data   Data which has not been divided into one or more variables, such as sex, age, smoking history, or any other category.

crude rate (Epidemiology)   The overall rate calculated from the actual number of events of interest, such as deaths in a total population during a specified period of time.

cryogenic liquid   A refrigerated, liquefied gas having a normal boiling point below $-130°F$ ($-90°C$).

cryogenics   The field of science dealing with the behavior of matter at very low temperatures.

crystal (Laser)   A three-dimensional atomic, ionic, or molecular structure consisting of periodically repeated, identically constructed, congruent unit cells.

crystalline material   A solid with an orderly array of atoms and molecules.

CSA   Canadian Standards Association.

CSHO (OSHA)   Compliance Safety and Health Officer.

CSP   Certified Safety Professional.

CT   Charcoal tube.

CTD   Cumulative trauma disorder.

CTGs   Control technology guidelines.

CTPV   Coal-tar pitch volatiles.

cubic centimeter   A volumetric measure equivalent to 1 milliliter.

cubic feet per hour   [cfh]   A measure of the volume of a substance flowing within one hour.

cubic feet per minute   [cfm]   A measure of the volume of a substance flowing within one minute.

cubital tunnel syndrome   Symptoms resulting from injury or compression of the ulnar nerve at the elbow, with pain and numbness of the hand and forearm, and weakness of the hand.

cu. ft.   cubic foot or cubic feet, ft^3.

cu. in.   cubic inch, in^3.

culture   The growing of microorganisms in a nutrient medium.

culture medium   Preparation suitable for the growth of cultures and cultivation of microorganisms.

cu. m.   cubic meter, m^3.

cumulative dose (Industrial Hygiene)   The average exposure of an individual to a contaminant over a period of time expressed as part per million–days/months/years. The most widely used expression of cumulative dose is part per million–years (ppm–yr. or ppm–y.). It is the product of the average concentration to which the person was exposed during the exposure period and the number of years of exposure.

cumulative dose (Ionizing Radiation)   The total radiation dose resulting from repeated exposure to ionizing radiation. It is expressed in rems or mrem.

cumulative effect (Toxicology)   The adverse health effect resulting from long-term exposure to a chronic toxicant.

cumulative trauma disorder   [CTD]   An occupational illness which develops over time and affects the musculoskeletal system and the peripheral nervous system as a result of repetitive motion of a body part, use of excessive force, or an awkward body position during work.

curie   [Ci]   A basic unit indicating the rate at which a radioactive material emits particles. One curie is equivalent to 3.7 E^{10} disintegrations or becquerels per second.

current   The flow of electrons through a conductor.

curettage   The removal of growths or other materials from the wall of a cavity or other surface.

curtain wall   The exterior part of a building, directly attached to the structure, extending from the roof to the ground.

cutaneous   Pertaining to or affecting the skin.

cutaneous hazard   Chemical which affects the dermal layer of the body.

cutie pie   A portable instrument used to determine the level of ionizing radiation.

cutting fluid   An oil–water emulsion that is used for cooling and lubricating the tool and work in machining and grinding operations.

cutting oil   An oil that is used for cooling and lubricating the tool and work in machining and grinding operations.

CV   Coefficient of variation.

CVS   Computer vision syndrome.

CWA   Clean Water Act.

C-weighted sound level   [dBC]   The sound level as determined on the C scale of a sound level meter or other noise survey meter with this weighting network.

CW laser   Continuous wave laser as opposed to a pulsed-type laser.

CWP   Coal workers' pneumoconiosis.

cyanosis   A bluish discoloration of the skin and mucous membranes due to an excessive concentration of reduced hemoglobin in the blood.

cycles per second (Acoustics)   [cps]   The number of complete wave changes that occur in one second. The use of cps has been replaced with the term *Hertz* (Hz).

cyclone   A size-selective device which is designed to separate coarse particulates from finer particles. In industrial hygiene sampling a cyclone is used to separate the respirable fraction of particulates in the sampled air from the total particulates drawn into the cyclone. The respirable particles are collected on a filter positioned downstream from the cyclone.

cyclotron   A particle accelerator which uses a magnetic field to confine a positive ion beam while an alternating electric field accelerates the ions in a spiral path.

cyst   An abnormal membranous sac containing a gaseous, liquid, or semisolid substance.

cystitis   Inflammation of the urinary bladder.

cytogenesis   The origin and development of cells.

cytogenetics   The branch of genetics devoted to the study of the cellular constituents concerned in heredity, i.e., chromosomes. The scientific study of the relationship between chromosomal aberrations and pathological conditions.

cytology   The branch of biology dealing with the study of the formation, structure, and function of cells.

cytopenia   A deficiency in the cellular elements of the blood.

cytoplasm   The protoplasm of a cell exclusive of the nucleus.

cytoscopy   The microscopic examination of cells for the purpose of diagnosing an illness.

cytotoxin   A toxin or antibody that has a specific toxic action upon cells of special organs. For example, a nephrotoxin would be a toxin that has specific destructive effect on kidney cells.

D

d   Deci, 1 E^{-1}.

d   Density.

da   Deca, 1 E^{1}.

damage   The severity of injury, or the physical, functional, or monetary loss that results when control of a hazard is not effective. Impairment of the usefulness or value of a person or property.

damage risk criterion (Acoustics)   The recommended baseline of noise tolerance that should not result in hearing loss.

damp (Vibration)   To reduce vibration by converting the vibration energy into heat.

damper   An adjustable source of airflow resistance, often installed at a right angle to air/gas flow, that serves as a means to regulate or distribute airflow in a ventilation system.

damping (Vibration)   Any means of dissipating vibrational energy within a vibrating medium. Damping converts mechanical energy to other forms of energy, usually heat.

dander   Small scales from the hair or feathers of animals which may cause allergic reactions in sensitive persons.

danger   The relative risk of exposure to a hazard. For example, a hazardous material may be present/used, but there may be little danger because of precautions taken to prevent exposure to it.

dark field (Microscopy)   A system of illumination in which the sample is illuminated with a hollow cone of light so that no direct light rays enter the objective lens, but only rays diverted by objects in the sample.

data   A collection of information, especially information organized for analysis or used as the basis for a decision.

data base   A file of records or a collection of data containing comparable information on different items and providing a means for organized information retrieval. Also referred to as a data bank.

daughter (Ionizing Radiation)   Synonym for the decay product of a radioactive material.

day–night sound level   The equivalent sound level over a 24-hour period. Adjustment is made for the sound level that occurs in the period from 10 P.M. to 7 A.M.

dB   Decibel.

dBA   Sound level in decibels as determined on the A scale of a sound-level meter. The A scale discriminates against low frequency sounds, as does the human ear; thus, the dBA scale is better for use in evaluating the hearing damage risk potential of a noise exposure.

dBB   Sound level in decibels as determined on the B scale of a sound-level meter.

dBC   Sound level in decibels as determined on the C scale of a sound-level meter. The dBC value approximates the overall noise level.

dc   Direct current.

DE   Desorption efficiency.

dead end corridor   An extension of a corridor or aisle beyond an exit or an access to an exit that forms a pocket in which occupants may be trapped.

dead hands   *See* Raynaud's syndrome.

dead-man switch   A fail-safe control mechanism that causes an operation/machine to stop if the operator should suffer a blackout or die at the controls.

dead room   A room that is characterized by an unusually large amount of sound absorption.

dead time (Instrument)   The interval of time between the instant of introducing a sample into the instrument to the first indication of response. Also referred to as lag time.

deaf   Term describing a person who has partially or completely lost the sense of hearing.

death certificate   A vital record, signed by a licensed physician or other designated health official, that includes the name, sex, birth date, date and place of death, cause of death, and whether the deceased was medically attended before death.

death rate   The ratio of the total deaths to the total population.

debilitating   Making feeble or abnormally weak.

debridement   The surgical excision of dead and devitalized tissue and the removal of all foreign matter from a wound.

deca   [da]   Prefix indicating 1 E^1.

decant   To pour off without disturbing the sediment.

decay   To decompose.

decay (Radioactive)   The disintegration of an unstable nuclide by the spontaneous emission of charged particles and/or photons. Indicates the decrease in the activity of a radioactive substance.

decay chain (Radioactive) A succession of radionuclides or daughter products that occur in the decay of a radioactive material until a stable nuclide is produced.

decay constant (Ionizing Radiation) The fraction of the number of atoms of a radionuclide which decay in a specific time period. Expresses the rate at which radioactive materials decay. Also referred to as the disintegration constant.

decay curve A graph showing the decreasing radioactivity of a radioactive source as time passes.

decay product The nuclide or daughter product resulting from the radioactive disintegration or decay of a radionuclide.

decay scheme The transformation of a radionuclide into its daughters, shown schematically, along with the percent decay by alpha, beta, or gamma emission and the energy (MeV) of the particle/photon.

deci [d] Prefix indicating $1 E^{-1}$.

decibel (Acoustics) [dB] A unit expressing the sound power level, which is the ratio of two amounts of acoustic signal power. It is equal to 10 times the logarithm to the base 10 of the ratio of a measured value to a reference value. One-tenth of a bel.

deciliter One-tenth of a liter, $1 E^{-1} L$.

decipol A unit for judging the perceived quality of outdoor air.

decision tree The alternative choices and possible outcomes expressed in quantitative terms that are available at each stage of the process of thinking through a problem and are arranged like the branches of a tree.

declination The angle, variable with location, between the direction in which a magnetic needle (compass) points and the true north-south line.

decompose To break down, separate, or resolve a compound into simpler compounds or basic elements.

decomposition The breakdown of a chemical or other substance by physical, chemical, biological, or other means.

decomposition product The material generated or produced by the physical or chemical degradation of a parent material.

decompression The process of gradually lowering elevated ambient pressure to eliminate dissolved gases from a diver's bloodstream and tissues.

decompression sickness A disorder characterized by joint pains, respiratory manifestations, skin lesions, and neurologic signs, occurring in aviators flying at high altitude and following rapid reduction of air pressure in persons who have been breathing compressed air in diving operations or in caisson work. *See* also bends.

decontaminate To remove a hazardous material or unwanted contaminant.

decontamination The process or effort to remove a contaminant from an individual, object, surface, material, or area to the extent necessary to preclude the occurrence of a foreseeable adverse health effect. The removal of a hazardous substance to prevent the occurrence of an adverse health effect that may result from exposure to it.

deflagration The thermal decomposition that proceeds at less than sonic velocity and may or may not develop hazardous pressures.

defoliant   A chemical sprayed or dusted on plants to cause the leaves to fall off.

degradation (Protective Clothing)   The loss in physical properties of an item of protective clothing due to exposure to chemicals, use, or ambient conditions (e.g., sunlight).

degradation (Respirator Filter)   A lowering of filter efficiency or a reduction in the ability of the filter to remove particles as a result of workplace exposure.

degreasing   The use of a solvent for the removal of grease, oil, or fatty materials from metal surfaces.

degree day   A unit used to estimate heating and cooling costs. For example, on a day when the mean temperature is less than 65°F, there are the same number of degree days as the mean temperature of the day that is below 65°F; that is, it is equivalent to the difference between a 65°F-base and the daily mean temperature. The degree days are used to estimate energy requirements for heating.

degrees of freedom   The number of independent comparisons that can be made between the members of a sample.

dehumidification   The process of removing moisture from air.

dehydration   The excessive loss of water from the body or from an organ or bodily part. The process of removing water from a substance or compound.

deionizing   The removal of dissolved salts from water by passing it through a deionizing column.

delamination   The separation of one layer of a material from another.

delayed effects (Toxicology)   Health effects which occur weeks, months, or years after a first exposure to a toxic substance.

delayed effects (Radiation Biology)   Those effects caused by ionizing radiation that do not become evident until years after exposure to ionizing radiation. Possible delayed effects are cancer in the exposed persons and genetic effects in their offspring.

deleading   The act of testing, removing, covering, containing, and disposing of paint, plaster or other material containing a dangerous level of lead paint.

deleterious   Having a harmful effect.

deliquesce   To absorb atmospheric water vapor and become liquid. Refers to water-soluble salts that dissolve in water absorbed from the air.

deluge shower   A shower unit which enables the user to have water cascading over the entire body. A minimum flow rate of water and time of use is recommended for effective contaminant removal using a deluge shower.

demand airline device   Respirator in which air enters the facepiece only when the wearer inhales.

demand respirator   An atmosphere-supplying respirator that admits breathing air to the facepiece only when a negative pressure is created inside the facepiece by inhalation.

dementia   Mental deterioration or a general decline or loss of intellectual ability.

de minimis   Not sufficient to be considered. The doctrine of law which states that the law does not deal with or take notice of insigificant matters.

demography   A branch of sociology or anthropology dealing with the study of the characteristics of human populations, such as size, growth, density, distribution, incidence of disease, vital statistics, etc.

demulsify   To resolve or break an emulsion, such as water and oil, into its components.

demyelination   The destruction of the myelin sheath of a nerve or nerves resulting in impairment of nerve function.

density   The ratio of the mass of a specimen of a substance to the volume of the specimen. The mass of a unit volume of a substance.

denuder   A device used to remove a gaseous contaminant from sampled air when monitoring for a substance(s) with which the denuded material would interfere.

deoxyribonucleic acid   [DNA]   The genetic material within the cell.

DEP   Department of Environmental Protection.

Department of Labor   [DOL]   A federal department that is responsible for improving working conditions of employees in industry in the U.S. and its possessions.

Department of Transportation   [DOT]   The federal agency that is responsible for establishing the nation's overall transportation policy.

dependent variable   A variable whose value is dependent on the effect of other variables.

depilatory   A material that has the ability to remove hair.

depleted uranium   Uranium having a smaller percentage of uranium 235 than that found in uranium as it occurs naturally.

depose (Law)   To declare under oath.

deposition (Aerosol)   Matter that settles out of the air onto a surface.

deposition (Legal)   Testimony under oath, especially in writing, by a witness for use in court in his/her absence.

depreciation   The decrease in the value of property due to normal wear or the passage of time.

depressant   A substance that diminishes bodily functions or activity.

DeQuervain's disease   A cumulative trauma disorder of the tendons of the wrists.

dermal   Pertaining to the skin.

dermal exposure   Being exposed to a chemical through skin contact.

dermal toxicity   The ability of a substance to produce a toxic effect by skin contact.

dermatitis   Inflammation of the skin from any cause. There are two general types of dermatitis: primary irritation dermatitis and sensitization dermatitis.

dermatosis   A broader term than dermatitis which includes any skin disease, especially those not characterized by inflammation.

dermis   The skin. *See* corium.

desalinization   The removal of salt from sea water to make it potable.

descriptive statistics   The collection, tabulation, and analysis of data in such a manner as to yield measures (i.e., mean, variance, standard deviation, etc.) that describe the population, group, or sample data.

descriptive study (Epidemiology) A study which describes the pattern of a disease, its symptoms, physiologic variables, and any health condition in one or more groups or populations.

desiccant A material that absorbs moisture from the air.

desiccate To make thoroughly dry.

designated person (OSHA) *See* authorized person.

desorption The process of removing an adsorbed material from the solid on which it is adsorbed and retained.

desorption efficiency The fraction of a known quantity of analyte that is recovered from a spiked solid sorbent media blank.

desquamation The sloughing off of the epidermal layer of the skin.

desulfurization The removal of sulfur and sulfur compounds from oil products.

detection limit The lowest concentration which can be determined that is statistically different from a blank sample. *See* also limit of detection.

detector (General) The portion of an instrument that is responsive to the material being measured.

detector (Ionizing Radiation) A device which converts ionizing radiation energy to a form more amenable to measurement. For example, an ionization detector, a scintillation detector, etc.

detector tube A sampling device/method for determining the concentration of specific contaminants in the air based on the reaction of the analyte (contaminant) with a reagent adsorbed on a sorbent. The length of color change produced in the tube or the intensity of the color produced is proportional to the analyte concentration and the sample volume.

determinate errors Errors which occur that are correctable if their causes can be determined.

detonation Thermal decomposition that occurs at supersonic velocity and is accompanied by a shock wave in the decomposing material.

detoxify To eliminate the toxic properties of a material through treatment.

developmental toxic effect Harmful effect to the embryo or fetus, such as embryotoxicity, fetotoxicity, or teratogenicity.

deviation The amount by which a score or other measure differs from the mean or other descriptive statistic.

dew The condensing of water vapor on cool surfaces, especially at night.

dew point The temperature at which the partial pressure of a vapor in a gas is equal to the saturation pressure. Temperature at which condensation occurs. Dew point is also commonly expressed as parts per million by volume.

DHEW Department of Health, Education, and Welfare.

diagnosis The process of determining health status and the factors for producing an observed effect.

dialysis A process by which various substances in solution with widely varying molecular weights may be separated by diffusion through semipermeable membranes.

diaphoresis   Sweating, especially profuse sweating.

diathermy   The therapeutic generation of local heat in body tissues by high-frequency electromagnetic waves.

diatomaceous earth   Soft, bulky solid material composed of the skeletons of small prehistoric aquatic plants related to algae. Natural diatomaceous earth is mostly amorphous silica, whereas calcined and flux-calcined diatomaceous earth can contain a significant amount of free silica as cristobalite.

diatoms   Microscopic, unicellular algae that are found on the sea floor and have siliceous cell walls.

dichotomous   Divided into two parts or classifications.

dielectric   A material that does not conduct electricity.

differential blood count   The determination of the number of whole blood cells in a cubic milliliter of blood.

differential diagnosis   The comparison of the symptoms of two or more similar diseases to determine which disease a worker has.

differential pressure   The difference between two fluid pressures, such as the difference in static pressure between two locations.

diffraction   Deviation of part of a beam of electromagnetic radiation.

diffraction (Acoustics)   The ability of a sound wave to pass around an obstruction with little loss of energy.

diffuse reflection   Different parts of a beam of light incident on a surface that are reflected over a wide range of angles.

diffuse sound field   A sound field in which the sound energy will flow in all directions with equal probability. This type of noise environment exists in a reverberation room and can be used to test sound-absorption materials.

diffuser   Component of a ventilation system that serves to distribute the air supplied to a space and promote air circulation throughout the area.

diffusion   The spontaneous mixing of one substance with another when in contact or separated by a permeable membrane or microporous barrier. In sampling it is the process by which an atmosphere being monitored is transported through a frit or membrane to a gas-sensing element by natural random molecular movement.

diffusion detector   A passive-type detection device which utilizes the principle of diffusion as the means to transport airborne contaminants to the detector. No mechanical means is employed to transport the sampled air from the surroundings to the detector. This is often referred to as a passive sampler, or passive sampling.

diffusion rate   The rate at which a gas or vapor disperses into or mixes with another vapor. A measure of the tendency of a gas or vapor to disperse into or mix with another gas or vapor.

diffusive sampling   *See* passive sampling.

dike   An embankment of earth and rock. A barrier to confine a material.

diligence   Persistent effort.

diluent   Material used to reduce the concentration of an active material to achieve a desirable effect.

dilute   To reduce the concentration of a material in the air or in a liquid.

dilution   The process of increasing the proportion of the solvent or diluent in a mixture.

dilution air   Uncontaminated air supplied from outside an area in which air is being exhausted to replace the air exhausted and dilute contaminants that may be generated in the space.

dilution ventilation   A ventilation system designed to provide airflow for maintaining an acceptable temperature or for diluting airborne contamination to an acceptable level. Contaminant concentration is reduced by mixing fresh, uncontaminated air with the contaminated air in the workspace. The object is to reduce and maintain below acceptable levels the concentration of small quantities of low toxicity contaminants that may be released constantly at some distance from exposed workers. This type of ventilation system is not recommended for health hazard control.

dioctyl phthalate   A colorless liquid that can be used to generate particles of uniform size (0.3 micrometers diameter) for use in testing the efficiency of filter media.

diopter   A measure of the power of a lens, equal to the reciprocal of its focal length in meters.

dip leg (Centrifugal Collector)   The pipe through which fines and exhaust air are transported out of the inverted cone collector where heavy particles are removed through the centrifugal action imparted to the air.

diplopia   The perception of two images of a single object. Also referred to as double vision.

direct ionizing particles   Electrically charged particles, such as electrons, protons, and alpha/beta particles, that have sufficient kinetic energy to produce ionization by collision.

direct-reading instruments   Instruments that provide an immediate indication of the concentration of airborne contaminants or physical agents in real time by direct readout on a meter, digital display, etc.

disability   Temporary or long-term reduction of a person's capacity to function.

disability benefits   Benefits that compensate a disabled worker for loss of income or earning capacity as a result of the disability.

disability days   The total number of days employees have been disabled during a year.

disabling injury   Bodily harm resulting in death, permanent disability, or any degree of temporary total disability. An injury which prevents a person from performing a regular established job for one full day beyond the day the injury occurred.

disabling injury frequency rate   The number of disabling or lost-time injuries per million employee-hours of exposure.

disabling injury severity rate   The total number of lost days per million employee-hours of exposure.

discipline   A branch of knowledge or learning, such as chemical engineering, physics, chemistry, industrial hygiene, etc.

disease   An abnormal condition of an organism or part, especially as a consequence of infection, inherent weakness, or environmental stress that impairs normal physiological function.

disinfect   To kill pathogenic microorganisms or their vectors in or on nonliving materials, or to inhibit their growth and render them inert.

disinfectant   An agent which renders pathogenic organisms inert and destroys infectious materials.

disinfection   The act or process of destroying organisms that may cause disease.

disinfestation   Any chemical or physical process which destroys or removes arthropods, rodents, or other small animal forms from clothing, person, domestic animals, or in the environment of persons or domestic animals.

disintegration (Ionizing Radioaction)   A spontaneous nuclear transformation characterized by the emission of energy and/or mass from the nucleus of an atom. The breaking up of an unstable atom.

disintegration constant   *See* decay constant.

disintegration rate (Ionizing Radiation)   The rate, expressed in half-life, at which radiation is emitted from a radioactive material.

dispersant   A chemical agent that is used to break up or disperse concentrations of a material, such as an oil spill in water.

dispersion (Atmosphere)   The mixing and movement of contaminants in their surroundings (e.g., air) with the resultant effect of diluting the contaminant.

dispersion (Fluids)   A suspension of fine particles in a liquid.

dispersion analysis (Air Pollution)   The determination, using a model, of the concentration of a contaminant at some location after its release, taking into account the physical and chemical properties and state of the substance, and the geographical, topographical, geological, and meteorological characteristics of the environment which influence the migration, dispersion, movement, or degradation of the substance in the environment.

dispersion staining   A particle identification technique in which the material of interest is immersed in a liquid media, such as an oil of specific index of refraction, and examined microscopically (e.g., by polarized light microscopy) for identification.

displacement (Vibration)   The change in distance or position of an object relative to a reference point.

disposable respirator   A respirator for which maintenance is not intended, and that is designed to be discarded after excessive resistance, sorbent exhaustion, physical damage, or at the end of service life when it is rendered unsuitable for use.

disposal   The discharge, dumping, injection, or placing of waste so that such waste, or constituent of the waste, may not enter the environment, be emitted into the air, or enter any waters.

dissipative muffler   A type of acoustic muffler that is typically used for reducing noise emissions from a source, such as a large engine. The muffler housing is lined with a sound-absorbing material.

dissociation   The separation of a molecule into two or more constituents as a result of added energy (e.g., heat) or the effect of a solvent on a dissolved polar compound.

dissolve   To cause a substance to go into solution.

dissolved oxygen   A measure of the amount of oxygen that is available for biochemical activity in a given amount of water.

dissolved solids (Water Quality)   Decomposed material contained in water.

distal phalanx   The last bony segment of a toe or finger. These bone structures were affected by VCM exposure, the diagnosis of which was referred to as acro-osteolysis.

distillate   The material removed in the distillation process.

distillation   The vaporization and subsequent condensing of a material in order to separate it from others.

distress   To cause anxiety or suffering.

diuresis   Excessive discharge of urine.

diuretic   A substance that promotes the excretion of urine.

diurnal   Pertaining to or occurring in a day or each day.

diving bell   A cylindrical or spherical compartment used to transport a diver(s) to an under-water work site.

dizziness   Having a sensation of disorientation, light-headedness, whirling, or of a feeling of a tendency to fall.

dL   Deciliter, $1 E^{-1}$ L.

DM respirator   Dust and mist respirator.

DMF respirator   Dust, mist, and fume respirator.

DNA   Deoxyribonucleic acid.

DOE   Department of Energy.

DOL   U.S. Department of Labor.

DOP   Dioctyl phthalate.

Doppler effect   The apparent upward shift in the frequency of a sound as the source approaches the listener, and the apparent downward shift in frequency as the source recedes.

DOS   Disk operating system.

dose (Industrial Hygiene)   The amount of material taken into the body or which the body received (e.g., physical agent) as a result of exposure to the hazard.

dose (Ionizing Radiation)   The amount of radiation energy absorbed by a specified area, volume, or the whole body, which may be expressed in rads, rems, or sieverts.

dose (Toxicology)   The amount of material given to an animal to determine if an effect is produced; or in many toxicity studies, the amount of material required to produce an effect.

dose–effect relationship   The relationship between the dose given and the occurrence and severity of the effect produced.

dose equivalent (Ionizing Radiation)   The product of the absorbed dose in rads and a factor (quality factor) that has been established for the various types of ionizing radiation (e.g., alpha, beta, gamma, neutron, etc.). Thus, it expresses all radiations on a common scale for calculating the effective absorbed dose.

dose rate (Industrial Hygiene)   The dose of a hazardous agent (chemical, physical, biological) delivered or taken into the body per unit time.

dose rate (Ionizing Radiation)   The absorbed ionizing radiation dose that is delivered to the specified area, volume, or whole body per unit of time.

dose ratemeter (Ionizing Radiation)   An instrument which measures ionizing radiation dose rate.

dose–response   Relationship between the dose received and the effect produced.

dose–response assessment   The determination of the relation between the magnitude of the exposure and the probability of the occurrence of an adverse health effect.

dose–response curve   A graphical representation of the response of an animal or individual to increasing doses of a substance.

dosimeter (Ionizing Radiation)   An instrument used to determine the radiation dose (i.e., from a gamma or an X-ray source) an individual has received. Often referred to as a pocket dosimeter.

dosimeter (Noise)   An instrument used to determine the time-weighted average noise level to which an individual was exposed. Some noise dosimeters can be used to determine additional noise exposure data.

dosing   The administration of some amount of material to an animal, organism, or other subject under study.

DOT   U.S. Department of Transportation.

double blind study   A study in which neither the participants nor the investigators are aware of who is receiving the material being studied or who is receiving the placebo.

double block and bleed   A method to isolate a piece of equipment, vessel, confined space, etc. from a line, duct, or pipe by locking or tagging closed two valves in series with each other in the line, duct, or pipe, and locking or tagging open to the outside atmosphere a bleed in the line between the two closed valves.

double-insulated tool   A tool having an insulation system comprised of the basic insulation both for the functioning of the tool and for basic protection against electric shock and supplementary insulation. The supplementary insulation is independent insulation provided in addition to the basic insulation to ensure protection against electric shock in case of failure of the basic insulation.

dpm   Disintegrations per minute.

draft (Ventilation)   The movement of air in a manner which results in discomfort to persons exposed to it due to its velocity, temperature, or other cause. It also refers to the difference in pressure between the inside and outside of a structure due to a combustion process (e.g., furnace, boiler, etc.). The draft causes the products of combustion to flow from the combustion process to the outside atmosphere. A backdraft can result if there is insufficient air for the combustion process.

Draize Test   An animal test procedure for assessing the potential irritation or corrosive effect of a material on the skin or eyes.

DRD (Ionizing Radiation)   Direct reading dosimeter (pocket type).

drift   The gradual and unintentional deviation of a given variable. Instrument drift is the gradual change in readout due to component aging, variation in power supply, characteristics of the detector, temperature effect on the detection system, etc.

droplet   Liquid particle that is suspended in air and that settles out quite rapidly.

droplet nuclei   Particles in the size range of 1–10 micrometers that represent the dried residue of droplets coughed or sneezed into the air.

drowsiness   The state of being half asleep.

drug interaction   A condition that develops when concurrent or nearly concurrent administration or exposure to two substances increases, decreases, or alters in some way the effect of one of the substances.

dry-bulb temperature   The air temperature as determined with a regular dry-bulb thermometer.

dry-bulb thermometer   An ordinary thermometer.

dry centrifugal collector (Ventilation)   *See* cyclone.

dry-gas meter   A secondary air flow calibration device, similar to a domestic gas meter, that can be used for determining the flow rate of air sampling pumps.

dry heat   A method of sterilizing by heating an object/container to 160°F/70 °C. This method requires a longer time for sterilization than does the wet-heat method.

dry ice   A trademark for solid carbon dioxide.

dry run   A trial run without the use of hazardous materials, or the operation of a process at less than design, in order to identify problems, verify operating characteristics/parameters, test procedures, etc.

duct   A conduit used for conveying air at low pressure.

ductless fume hood   A hood which returns filtered air to the area where it is located. This type hood is to be used only with nontoxic chemicals. Also referred to as a ductless lab hood.

ductless lab hood   *See* ductless fume hood.

duct silencer   Sound-absorbing material that is typically placed on the inside of ductwork to absorb sound and prevent it from being transmitted through the duct to occupied areas.

duct velocity   The air velocity through a duct cross section.

Dunn cell   A glass slide device formerly used to contain an aliquot of the dust-collecting media in which the airborne dust sample was collected and that enabled counting of the dust so that a determination of its concentration could be made.

duplicate analysis   A repeat analysis of a single sample.

duplicate samples   The collection of two samples at the same location and for the same period and sample rate in order to provide an indication of the reproducibility of the sampling and analytical method, as well as an indication of the uniformity of the atmosphere being sampled.

duration of exposure   The period of time during which exposure to a hazardous substance or physical agent occurs, or how long a time one works with a substance or in the environment where the agent is used.

dust   Small solid particles created by the break up of larger particles, such as by crushing, grinding, drilling, handling, detonation, impact, etc. Dusts in the industrial environment typically do not flocculate (join together) in air, but settle out under the influence of gravity.

dust collector   An air-cleaning device for removing particulates from air. Dust collectors are usually designed for higher dust loadings than those handled by ventilation systems that are typically equipped with filters.

dust explosion   A dust combustion process so confined as to result in an appreciable rise in pressure.

dust loading   The amount of dust in an exhaust gas expressed typically in grains per cubic foot.

dust mask   *See* filtering facepiece.

dustproof   Constructed such that dust will not interfere with its operation.

dust spot efficiency (Ventilation)   A measure of the efficiency of a filter media in a ventilation system.

duty cycle (Laser)   The fraction of time that a laser system is operated.

D-weighted noise level   Weighting level on some sound level meters for determining the offensiveness of aircraft noise.

dwt   Deadweight tons.

dynamic loss (Ventilation)   The loss of static pressure in ductwork due to turbulence, such as that which occurs when air enters or that caused by elbows and other fittings.

dyne   The force that gives a mass of one gram an acceleration of one centimeter per second per second.

dysarthria   Imperfect articulation of speech due to disturbances of muscular control that result from damage to the central nervous system or peripheral nervous system.

dysbarism   A general term applied to any clinical syndrome caused by a difference between the surrounding atmospheric pressure and the total gas pressure in the various tissues, fluids, and cavities of the body.

dyscrasia   A morbid condition caused by poisons in the blood.

dysentery   Term given to a number of disorders marked by inflammation of the intestines, especially of the colon, and attended by pain in the abdomen, with frequent stools containing blood and mucus.

dysfunction   A disturbance, impairment, or abnormality of the functioning of an organ.

dyspepsia   Impairment of the function of digestion.

dysphagia   Difficulty in swallowing.

dysplasia   Abnormal growth or development of organs or cells.

dyspnea   Difficult or labored breathing.

dystrophy   Any disorder caused by defective nutrition.

dysuria   Painful or difficult urination.

E

E   Exa, 1 E $^{-18}$.

E85   Motor fuel containing 15 percent gasoline and 85 percent ethyl alcohol.

ear defender   Devices, such as ear plugs, ear muffs, canal caps, etc. that are used by individuals to provide personal hearing protection from noise.

eardrum   A membrane in the ear canal that vibrates when sound waves enter the ear. Also called the tympanic membrane. It serves as a boundary between the external ear and the middle ear.

ear insert   A hearing protective device that is designed to be inserted into the ear canal in order to reduce the level of noise reaching the hearing-sensitive part of the ear.

ear muffs   Hearing protective device consisting of two interconnected acoustically treated cups that fit over the ears and that reduce the level of noise reaching the hearing-sensitive part of the ear.

earphone   An electroacoustic transducer intended to be closely coupled acoustically to the ear.

ear plugs   Hearing protection devices made of a nonporous, pliable material and that are designed to fit in the ear canal.

EC   European Community.

ecchymosis   A small hemorrhagic spot in the skin or mucous membrane forming a non-elevated blue or purplish spot.

ECG/EKG   Electrocardiogram.

ECD   Electron capture detector.

ECD   Emission control device.

ecology   The science of the relationships between living organisms (e.g., humans) and their physical environment.

economic poison   Chemical used to control pests or defoliate crops.

ecosphere   The layer of earth and troposphere inhabited by or suitable for the existence of living organisms.

ecosystem   The interactive system consisting of the living community and the nonliving surroundings. A group of plants and animals occurring together plus that part of the physical environment with which they interact.

ECRA   Environmental Cleanup Responsibility Act.

ectoderm   The outer layer of skin cells.

eczema   A noncontagious inflammation of the skin marked by redness, itching, and the outbreak of lesions that discharge serous matter and become scaly.

ED–50   effective dose–50 percent.

eddy   A current of water or air that is moving contrary to the direction of the main current.

edema   A local or generalized condition in which the body tissues contain an excessive amount of tissue fluid. Swelling of body tissue.

EDF   Environmental Defense Fund.

eductor   *See* ejector.

EEC   European Economic Community.

EEG   Electroencephalogram.

EEL   (OSHA)   Emergency Exposure Limit.

effect   The consequence, outcome, or result of a stimulus or action.

effective dose–50 percent   [ED–50]   The dose of a substance that will create a specific effect in 50 percent of the test animals.

effective exposure time   The minimum exposure time required to produce a health effect.

effective half-life (Radiation Biology)   The time required for a radionuclide to be reduced to one-half of its initial activity in the body as a result of the combined action of radioactive decay and biological elimination.

effective stack height   The sum of the actual stack height and the rise of the plume after emission from the stack.

effective temperature   The combination of dry-bulb and wet-bulb temperature of slowly moving air that produces like sensations of warmth and coolness. The combinations of dry-bulb and wet-bulb temperature and air movement are located on an effective temperature chart from which the effective temperature can be read.

effective temperature index   [ET]   An arbitrary index that combines into a single value the effect of temperature, humidity, and air movement on the sensation of warmth or cold felt by the human body. A sensory index, developed by ASHRAE, of the degree of warmth that a person, stripped to the waist and engaged in light activity, would experience upon exposure to different combinations of air temperature, humidity, and air movement. This index is applicable to work situations in which light activity is performed over a several-hour period. A revised effective temperature chart has been developed for sedentary type work situations, as well as one (CET) where radiant heat is a concern.

efferent   Directed away from a central organ, especially impulses from the central nervous system to activate a gland or muscle.

efficacy   The capacity or ability to produce the desired effect.

efficiency   Typically the ratio of output to input.

effluent   Waste liquid, gas, or vapor that results from an industrial process and that is discharged, treated or untreated, into the environment.

egress   The arrangement of an exit from a structure/building to ensure a safe means of exiting.

EH & S   Environmental health and safety.

EHS   Extremely hazardous substance or extraordinarily hazardous substance.

EIA   Environmental impact assessment.

EIS   Environmental impact statement.

ejector An air-moving device employing compressed air to create a vacuum as it is passed through a venturi or straight pipe, which then induces air to flow. Often used when contaminated air could corrode a fan if it were passed through it. Ejectors are not very efficient airmoving devices, but do have application in special situations. Sometimes referred to as educators.

elastomer A rubber or rubberlike material; for example, a synthetic polymer with rubber-like characteristics.

elbow (Ventilation) A fitting that allows two pipes to be joined together at an angle of less than 180 degrees, usually 90, 45, or 30 degrees.

electrical conductivity A measure of the ability of a solution to conduct electric current.

electrical ground A conducting connection between an electrical circuit and the earth or some conducting body which serves in place of the earth.

electrocardiogram [ECG/EKG] A graphic tracing of the electric current produced by the excitation of the heart muscle.

electrochemical detector A detector that operates on the principle of electrochemical oxidation or reduction of a specific chemical in an electrolyte or galvanic cell. The electrons produced in the chemical reaction are proportional to the contaminant concentration.

electrode A solid electric conductor through which an electric current enters or leaves a medium such as an electrolyte.

electroencephalogram [EEG] A graphic tracing of the electric current developed in the brain as detected by electrodes placed on the scalp.

electrolysis The process of conducting an electric current by means of a chemical solution.

electrolyte A chemical that, when dissolved in water, dissociates into positive and negative ions.

electromagnetic field (Physics) The field created by the interaction of an electric field and a magnetic field when an electric current passes through a wire.

electromagnetic interference [EMI] A disturbance in the operation of an electronic device that results from the presence of local electric and magnetic fields.

electromagnetic radiation Non-ionizing radiations in the range of wavelengths from $1 E^{-12}$ centimeters to greater than $1 E^{10}$ centimeters. Includes ultraviolet, visible, infrared, microwaves, and radio frequency waves.

electromagnetic spectrum The range of frequencies and wavelengths emitted by atomic systems. The spectrum includes radiowaves as well as the short cosmic rays.

electromagnetic susceptibility Degraded performance of an instrument caused by an electromagnetic field.

electromotive force [EMF] The electric voltage due to the difference in the potential between two dissimilar electrodes.

electron An elementary particle with the properties of both a particle and a wave, having a negative electric charge of $1.60210 E^{-19}$ coulombs, and existing as a constituent of an atom or in the free state (e.g., a beta particle).

electron capture (Ionizing Radiation)   The capture of an electron in the innermost orbit of an atom by the nucleus. Energy is emitted in the form of a gamma emission or internal conversion electrons. Electron capture and internal conversion are accompanied by the emission of X-rays.

electron capture detector   A type of detector employed in gas chromatography.

electron microscopy   An analytical method that utilizes a beam of electrons for the analysis of materials. This methodology is used for the identification of asbestos and other materials.

electron shells   The position of the electrons in an atom. The positions are designated by letters signifying decreasing energy levels moving away from the nucleus. The K ring is nearest the nucleus, followed by the L, M, N, etc. rings in sequence moving out from the nucleus.

electron volt   [eV]   A unit of energy equivalent to the energy gained by an electron passing through a potential difference of one volt. One electron volt is equal to $1.6 \, E^{19}$ joule.

electrophoresis   A technique for separating materials in suspension in a liquid through the influence of an applied electric field causing the charged particles to move.

electroscope   An instrument used to detect the presence, sign, and in some configurations, the magnitude of an electric charge by the mutual attraction or repulsion of metal foils.

electrostatic precipitator   [ESP]   A device used to remove particulates from an airstream by charging the particles when they are passed through an electric field and then collecting them on an electrode (plate) of opposite polarity.

element   A chemical substance which cannot be divided into simpler substances by chemical means. One of 106 presently known substances that comprise all matter at and above the atomic level.

ELF   Extremely low frequency range of radio frequency radiation (3 to 3000 hertz).

ELF EM field   Extremely low frequency electromagnetic field.

elimination (Toxicology)   The elimination of a substance from the body via expiration, urination, or defecation.

eluent   The washings in the separation of a material by elutriation.

elutriation   The process of separating lighter particles from heavier ones by passing a fluid stream (e.g., air or water) upward through the mixture. The heavy materials settle to the bottom, and the lighter materials remain suspended and, thus, are at the top.

elutriator   A sampling device which separates particles according to mass and aerodynamic size by maintaining laminar flow through it, thereby permitting particles of greater mass to settle out rapidly with the smaller particles depositing at greater distances from the entry point of the elutriator.

emaciated   Extremely lean, wasted, or feeble condition of the body.

embolism   The sudden blockage or closure of an artery or vein due to a blood clot or foreign material.

embryo/fetus   The developing organism from conception until the time of birth.

embryotoxicity   The toxic effect of a substance on the embryo.

embryotoxin   A material that is harmful to the developing embryo. Substances that act during pregnancy to cause adverse effects on the fetus.

EMCS   Energy Management and Control System.

emergency   Any occurrence or event that could endanger workers or equipment, such as equipment failure, the rupture of a container, failure of control equipment, or any other event that would result in a hazardous situation.

emergency action plan   A plan that describes the procedures the employer and employees must take to ensure employee safety from fire or other emergencies.

emergency exposure limit   [EEL]   The concentration of an air contaminant to which, it is believed, an individual can be exposed in an emergency without experiencing permanent adverse health effects, but not necessarily without experiencing temporary discomfort or other evidence of irritation or intoxication.

emergency lighting   A system for providing adequate illumination automatically in the event of interruption of the normal lighting system. The emergency lighting should provide, throughout a means of egress, not less than one foot-candle of illumination for a period of one and one-half hours.

emergency medical technician   [EMT]   A person certified by the Emergency Medical Services Commission who is responsible for the administration of emergency care to emergency patients as well as for their handling and transportation.

Emergency Planning and Right to Know Act   [EPCRA]   An act that requires the reporting of spills that occur outside a facility and impact the environment.

emergency procedure   An action plan to be implemented in the event of an emergency.

emergency respirator use   The use of a respirator when a hazardous atmosphere develops suddenly and requires its immediate use for escape or for responding to an emergency in locations/areas/operations where a hazardous situation may exist or arise.

emergency response (HAZWOPER)   A response effort to an occurrence which results, or is likely to result, in an uncontrolled release of a hazardous substance.

emergency response personnel   Individuals authorized to respond to fires, rescue, and hazardous material emergencies within a specified jurisdiction.

Emergency Response Planning Guides   Concentration ranges, developed by an AIHA committee, above which adverse health effects could reasonably be expected to occur if exposures exceed the time limit established for the guides. Different effects are identified for exposure periods of one hour in ERPG-1, ERPG-2, and ERPG-3.

emergency response team   Employees of a facility who have been designated by management to be responsible for implementing and carrying out the sites emergency response plan in the event of an emergency, including the release of a hazardous substance.

emergency shower   A water shower designed and located for use if an employee or other individual contacts a material that must be removed promptly in order to prevent an adverse health effect. A recommended flow rate and application period have been established for its effective use. Also referred to as a drench shower.

emergency situation (OSHA)   Any occurrence such as, but not limited to, equipment failure, rupture of a container, or failure of control equipment that may, or does, result in an uncontrolled significant release of an airborne contaminant.

emesis   The act of vomiting.

emetic   An agent which induces vomiting.

emf   Electromotive force.

emf   Electromagnetic force.

EMF   Electromagnetic field.

EMF   Electric and magnetic field.

EMI   Electromagnetic interference.

emission (Air/Water)   The release or discharge of a substance or stream into the atmosphere or receiving body of water.

emission control device   [ECD]   Equipment used for reducing the emission of pollutants.

emission inventory   A listing of air pollutants emitted into the atmosphere, such as pounds per day of a particular substance, by type of source.

emission rate   The amount of pollutant emitted per unit of time.

emissions   The uncontrolled release of gases, vapors and/or aerosols into the environment, including the work area, as well as the into the ambient atmosphere.

emission standard   The amount of a pollutant permitted to be discharged from a source.

emissivity   The ratio of the radiation intensity from a surface to the radiation intensity of the same wavelength from a black body at the same temperature. The emissivity of a perfect black body is 1.

emphysema   Overdistention of the alveolar sacs of the lungs. A condition of the lungs in which there is dilation of the air sacs, resulting in labored breathing and increased susceptibility to infection.

empirical   Derived from practical experience or relying on observations or experimental results as opposed to theory.

employee exposure (Respiratory Protection)   Exposure to a concentration of an airborne contaminant that would occur if the individual were not using respiratory protection.

employee exposure record   A record containing environmental monitoring data or measurements of a toxic substance or harmful physical agent; biological monitoring results that directly assess the absorption of a toxic substance or harmful physical agent by body systems; and any records which reveal where and when toxic substances are used or when there is exposure to a harmful physical agent.

employee medical record (OSHA)   A record concerning the health status of an employee, which is made or maintained by a physician, nurse, other health professional or technician.

EMS   Environmental management system.

EMT   Emergency medical technician.

emulsion   A mixture of one liquid that is distributed uniformly in small globules throughout a second liquid. Types include oil-in-water and water-in-oil.

encapsulant   A material that can be applied to a solid or semi-solid material to prevent the release of a component(s), such as fibers from an ACM.

encapsulate   To seal a substance in an impervious material which will not be degraded physically or chemically.

encapsulation The coating of an asbestos-containing material, man-made mineral fiber, lead-containing material, or other material from which release of a contaminant is to be controlled by the encapsulating material. An example is the coating of asbestos-containing material with a bonding or sealing agent to prevent the release of fibers.

encephalitis Inflammation of the brain.

encephalopathy Any degenerative disease of the brain.

enclosing hood An exhaust hood which partially or totally encloses a source of contamination and is often provided with exhaust ventilation.

enclosure Mechanically fixing a material around a substance, machine, surface, etc. to reduce the release or abrading of a hazardous agent into the work or living environment.

enclosure (Asbestos) A tight structure around an area of asbestos-containing material designed to prevent the release of fibers into the surrounding area.

endangered species Any species of living organism faced with the likelihood of extinction.

endemic The usual frequency of a disease occurrence. The continuing prevalence of a disease among a population or in an area.

endemic disease The constant presence of a disease or infectious agent within a given geographic area.

endocrine disrupter A chemical substance or mixture that alters the structure or function of the endocrine system and causes adverse health effects.

endocrine system The system of glands and other structures that regulate the secretions of hormones directly into the circulatory system and that influence metabolism and other body processes. Organs having endocrine function include the thyroid, parathyroid, pituitary and adrenal glands, the pancreas and others.

end-of-service-life indicator A method of warning a respirator user of the approach of the end of adequate protection by the respiratory device.

endogenous Originating within an organ or part.

endoscopy The visual examination of organs and spaces within the body using an endoscope.

endothermic Characterized by heat absorption. A reaction in which heat is absorbed is called an endothermic reaction.

endotoxin A toxin that is present in the bacterial cell but not in cell-free filtrates of cultures of intact bacteria.

end point (Toxicology) A specific measurable health or environmental effect used to assess the toxicity of a substance.

energy The capacity for doing work. The product of power (watts) and time duration (seconds) where one watt-second equals one joule. Forms of energy include chemical, nuclear, kinetic, and others.

Energy Management and Control System An energy management control system for managing a building's use of energy for heating, ventilating, air conditioning, lighting, and/or business related processes, as well as for fire control, safety, and security.

Energy Research and Development Administration [ERDA] That part of the AEC that became the reactor development section and was subsequently incorporated into the Department of Energy.

engineering control   Any procedure other than administrative control or personal protection that reduces exposure to an airborne contaminant, physical agent, or other health hazard. Method of controlling exposure to an air borne contaminant or physical stress by modifying the source, reducing the amount of contaminant or physical agent released into the work environment, or preventing the contaminant or physical agent from reaching those potentially exposed to it.

engulfment (Confined Space)   The surrounding and effective capture of a person by a liquid or finely divided (flowable) solid substance that can be aspirated to cause death by filling or plugging the respiratory system or that can cause enough force on the body to result in death by strangulation, constriction, or crushing.

enriched material   A material in which the relative abundance of the isotopes of a given element is altered, thus producing a form of the element that has been enriched in one particular isotope and depleted in its other isotopic forms.

enteric   Pertaining to the small intestine.

enteritis   Inflammation of the intestines, particularly the small intestine.

enterotoxin   A toxin specific for cells of the intestinal mucosa. On ingestion it can produce violent vomiting and diarrhea and is the primary factor in staphylococcal food poisoning.

enthalpy   Heat function at constant pressure. Enthalpy is sometimes also called the heat content of the system.

entomology   The scientific study of insects.

entrainment   A process in which suspended droplets of liquid, particles, or gas are drawn into the flow of a fluid and transported.

entropy   A measure of the degree of disorder in a system, wherein every change that occurs and results in an increase of disorder is said to be a positive change in entropy. All spontaneous processes are accompanied by an increase in entropy. The internal energy of a substance that is attributed to the internal motion of the molecules.

entry (Confined Space)   The act of passing through an opening into a confined space; the ensuing work in the space. An entry occurs when any part of the body breaks the plane of an opening of what is classified as a confined space. An alternate definition is any action resulting in any part of the face of the employee breaking the plane of any opening of a confined space, as well as any ensuing work inside the space.

entry loss   Loss in pressure caused by air flowing into a duct or hood opening.

entry permit   The written authorization of the employer for entry into a confined space under defined conditions for a stated purpose during a specified time.

entry permit (OSHA)   A written or printed document that is provided by the employer to allow and control entry into a permit space and that contains required information (e.g., space to be entered, purpose of the entry, date and duration of permit, names of authorized entrants and attendant, etc.).

enuresis   Involuntary urination.

environment   The water, air, and combinations of physical conditions that affect and influence the growth and development of organisms. The surrounding conditions to which an employee is exposed.

environmental chemistry   The study of the sources, reactions, transport, effects, and fates of chemical species in air, soil, and water environments, and the effects of technology on them.

environmental factors (Indoor Air Quality)   Conditions, other than indoor air contaminants, that cause stress, discomfort, and/or health problems.

environmental fate   The fate of a chemical or biological pollutant following its release into the environment. Factors affecting this fate include time, temperature, humidity, sunlight, and the presence of other chemical substances.

environmental health   The activities necessary to ensure that the health of employees, customers, and the public is adequately protected from any health hazards associated with a company's operations.

environmental impact assessment   A report prepared by an applicant for a discharge permit that identifies and analyzes the impact of a new source of emission to the environment and discusses possible alternatives.

environmental impact statement   [EIS]   A document required by the federal agencies for major projects which potentially will release contaminants into the environment. The document provides information on the effects of a release and alternatives for carrying out the project as proposed.

environmental lapse rate   The distribution of the temperature vertically. Also called the lapse rate.

environmental monitoring   The systematic collection, analysis, and evaluation of environmental samples, such as from air, to determine the contaminant levels to which workers are exposed.

environmental noise   The sound intensity and the characteristics of sounds from all sources in the surrounding environment.

Environmental Performance Evaluation   [EPE]   A system to measure an organization's environmental progress against the ISO 14,000 program goals.

Environmental Protection Agency   [EPA]   An agency of the federal government whose objective is to protect and enhance our environment today and for the future. It is responsible for pollution control and abatement, including programs for air, water pollution, solid and toxic waste, pesticide control, noise abatement, and other pollution sources and concerns.

environmental science   The science of the complex interactions that occur among the terrestrial, atmospheric, aquatic, living, and anthropological environments.

environmental tobacco smoke   [ETS]   Contamination of the air in the breathing zone of nonsmokers that poses a risk to their health and is a result of persons smoking in the area/room/building.

environmental toxicology   The study of the potential effects of contaminants in the environment as they relate to individuals affected by exposure, time, dose, and biological effects.

environmental transformation   The conversion of a chemical in the environment from one form to another.

enzyme   A biological catalyst which generally activates or accelerates a biochemical reaction.

EOE   Equal opportunity employer.

EP   Extraction Procedure Toxicity Test.

EPA   U.S. Environmental Protection Agency.

EPCRA   Emergency Planning and Community Right to Know Act.

EPE   Environmental Performance Evaluation.

epicondylitis   Inflammation or infection in the general area of the elbow, such as tennis elbow.

epidemic   Unusually frequent occurrence of disease in the light of past experience in the same geographical area.

epidemiology   The study of the frequency of occurrence and distribution of a disease throughout a population, often with the purpose of determining the cause. To the industrial hygienist, it is the determination of statistically significant relationships of specific diseases of specific organs of the human body in selected occupational groups (cohorts) in comparison with selected controls.

epidermis   The outer, protective, nonvascular layer of the skin.

epigenetics   The science concerned with the causal analysis of development of an organism.

epilation   The removal of hair by the roots. Loss of body hair.

epileptiform   Resembling epilepsy or its manifestations.

episode (Air Pollution)   An incident within a given region as a result of a significant concentration of an air pollutant with meteorological conditions such that the concentration may persist and possibly increase, with the likelihood that there will be a significant increase in illnesses and possibly deaths, particularly among those who have a preexisting condition that may be aggravated by the pollutant.

episode (Epidemiology)   The period in which a health problem or illness exists, from its onset to its resolution.

epistaxis   A nosebleed.

epithelial   Pertaining to or comprised of epithelium.

epithelial cells   Cells that make up the tissues covering the skin and various organs of the body.

epithelioma   Tumor derived from epithelium.

epithelium   The covering of the internal and external surfaces of the body, including the lining of vessels and other small cavities.

epizootic   A disease epidemic in nonhuman animals.

EPRI   Electric Power Research Institute.

EPTOX   Extraction Procedure Toxicity Test.

EP toxicity test   A test to determine the presence and concentration of specific toxic substances.

EP toxic waste   A waste with certain toxic substances present at levels greater than limits specified by regulation.

equal opportunity employer   An employer who hires employees without regard to race, sex, or ethnicity.

equilibrium   A state of balance.

equivalent diameter   *See* aerodynamic diameter.

equivalent human dose (Toxicology)   An animal dose that has been adjusted to its effect in humans by applying scaling factors.

equivalent method   A method for sampling and analyzing air samples which has been demonstrated to have a consistent and quantitatively equivalent relationship to a reference method under specified conditions.

equivalent sound level   [Leq]   The sound level in decibels of the mean square A-weighted sound pressure to which a person was exposed over a stated time period.

equivalent weight   The weight of an element that combines chemically with 8 grams of oxygen or its equivalent.

eradicate   To eliminate or destroy totally.

eradication of a disease   The termination of all transmission of infection by extermination of the infectious agent through surveillance and containment measures.

ERDA   Energy Research and Development Administration.

erg   A force of 1 dyne acting through a distance of 1 centimeter.

ergonomics   The application of human biological sciences in conjunction with engineering disciplines to obtain benefits of better efficiency and the well-being of workers through the study of people in their work environment, especially of their interactions with machines and tools. Ergonomics has often been referred to as the effort to fit the job to the person and not the other way around.

ergonomist   An individual trained in health, behavioral, and technological sciences, and who is competent to apply those fields to the industrial environment to reduce stress on personnel, thereby preventing work strain from developing to pathological levels or producing fatigue, careless workmanship, or high employee turnover.

ERMAC   Electromagnetic Radiation Management Advisory Council.

erode   To deteriorate or wear away.

erosion   A process in which a material, such as soil, is worn away or removed by the action of wind or water.

ERPG   Emergency Response Planning Guideline.

error   The difference between the true or actual value to be measured and the value indicated by the measuring system. Any deviation of an observed value from the true value.

erysipelas   A contagious disease of the skin and subcutaneous tissue caused by streptococcus and marked by redness and swelling of affected areas.

erysipeloid   A bacterial infection affecting slaughterhouse workers and fish handlers.

erythema   Reddening of the skin. Abnormal redness of the skin due to distention of the capillaries with blood.

erythemal region (Electromagnetic Radiation)   The electromagnetic spectrum in the ultraviolet region from 2800 angstroms to 3200 angstroms.

erythrocytes   Components of peripheral blood that contain hemoglobin and that are referred to as red blood cells.

escape-only respirator   A respirator intended to be used only for emergency exit.

eschar   Damage created to the skin and underlying tissue from a burn or as a result of contact with a corrosive material.

ESLI (Respirator)   End-of-service-live indicator.

ESP   Electrostatic precipitator.

ET   Effective temperature.

ethics   The study of morality, moral thought, and moral conduct.

etiologic agent   The agent that causes a disease or adverse health effect.

etiology   Cause. The study or theory of the factors that cause disease and their methods of introduction to the host.

ETOH   Ethyl alcohol.

ETS   Environmental tobacco smoke.

ETS (OSHA)   Emergency temporary standard.

eugenics   The study of hereditary improvement by genetic control.

euphoria   The absence of pain or distress. An exaggerated sense of well-being.

eupnea   Normal quiet breathing.

eustacian tube   A cartilagenous tube connecting the tympanic cavity with the nasopharynx; a tubular structure leading from the back of the throat to the middle ear.

eutrophic   Term designating a body of water in which an increase in mineral and organic nutrients has reduced the dissolved oxygen, producing an environment that favors plant over animal life.

eutrophication   The slow process of the aging of a lake and its evolving into a marsh, eventually disappearing due to plant growth.

EUV   Extreme ultraviolet.

eV   Electron volt.

evaporation   The change of a substance from the solid or liquid phase to the gaseous or vapor phase.

evaporation rate   The ratio of the time required to evaporate a measured amount of a liquid to the time required to evaporate the same amount of a reference liquid under ideal test conditions. Normal butyl acetate has typically been used as the reference standard.

evase   A gradual enlargement at the outlet of an exhaust system to reduce the air discharge velocity efficiently so that velocity pressure can be regained instead of being wasted as occurs when air is discharged directly from a fan housing.

exa   [E]   Prefix indicating 1 E^{18}.

exacerbate   To increase the severity.

exacerbation (Epidemiology)   An increase in the severity of a disease or any of its symptoms.

exceedance (Air Pollution)   Violation of an environmental protection standard by exceeding an allowable limit or concentration.

excess air   A quantity of air in excess of the theoretical amount required to completely combust a material, such as a fuel, waste, etc. Also referred to as excess combustion air and expressed as a percentage (e.g., 20 percent excess air).

excess flow valve   A type of safety valve designed to shut off if the flow rate exceeds a set value.

excise   To remove by cutting out.

excitation   The addition of energy to a system, thereby transferring it from its ground state to an excited state.

excited state   An atom with an electron at a higher energy level than it normally occupies. This principle is employed in the use of TLDs for determining exposure to ionizing radiation with this type device.

excreta   Waste material excreted from the body, including urine and feces.

excursion   A movement or deviation from the norm. In industrial hygiene, it is the deviation above the norm that is of concern.

excursion limit   The amount by which an exposure limit can be exceeded, and the number of times in an exposure period it can be exceeded without causing an adverse health effect, narcosis, or discomfort, impairing self rescue, or reducing work efficiency.

exfiltration   The flow of air from inside a building to the outside due to the existence of negative pressures outside the building surface.

exfoliation   The peeling or flaking off of the skin.

exhalation valve (Respiratory Protection)   A device which permits exhaled air to be discharged from a respirator and prevents outside air from entering.

exhaust air   Air removed from an enclosed space and discharged to the atmosphere.

exhaust grill   Fixture in the wall, floor, or ceiling through which air is exhausted from a space.

exhaust hood   A structure used to enclose or partially enclose a contaminant-producing operation or process, or used to guide air flow in an advantageous manner to capture a contaminant, and that is connected to a duct/pipe or channel designed to remove the contaminant from the hood.

exhaust rate   The volumetric flow rate at which air is removed by a ventilation system.

exhaust system   A system for removing contaminated air from a space, comprised of two or more of the following elements: an exhaust hood, duct work, air-cleaning equipment, exhauster, or discharge stack.

exhaust ventilation   Mechanical removal of air from a portion of a building or other area/space.

eximer   A gas mixture used as the active medium for lasers emitting ultraviolet light.

exit   That portion of a means of egress that is separated from all other spaces of the building or structure by construction or equipment to provide a protected way to travel to the exit discharge.

exit access   That portion of a means of egress that leads to an exit.

exit discharge   That portion of a means of egress between the termination of an exit and a public way.

exogenous   Derived or developed from external causes.

exothermic   Characterized by evolution of heat. A reaction in which heat is given off is called an exothermic reaction.

exotoxin   A toxin excreted by a microorganism into a surrounding medium.

expected deaths (Epidemiology)   The estimate of the number of deaths that result from a specific disease or cause.

expectorate   To cough up and eject from the mouth by spitting.

expertise   Specialized knowledge.

expiration   The act of breathing out.

expiration date (Detector Tubes)   The date beyond which colorimetric detector tubes should not be used.

expiratory flow rate   The maximum rate at which air can be expelled from the lungs.

explosimeter   A device for detecting the presence of and measuring the concentration of gases or vapors that can reach explosive concentrations.

explosion   A rapid increase of pressure, followed by its sudden release and expansion of gases.

explosion-proof   The design of a device or equipment to eliminate the possibility of its igniting volatile material. A type of construction that is designed to contain an explosion and prevent its propagation to the atmosphere outside the device/equipment.

explosion-proof apparatus   Apparatus enclosed in a case that is capable of withstanding an explosion of a specified gas or vapor that may occur within it; and that is capable of preventing the ignition of a specified gas or vapor surrounding the enclosure by sparks, flashes, or explosion of the gas or vapor within; and that operates at such an external temperature that a surrounding flammable atmosphere will not be ignited by it.

explosive   Any chemical compound, mixture, or device, the primary or common purpose of which is to function with substantially instantaneous release of gas and heat.

explosive atmosphere   An atmosphere containing a mixture of vapors or gases that is within the explosive or flammable range. Also referred to as an explosive mixture.

explosive limit   *See* flammable limit.

exponent   Any number or symbol to the right and above another number, symbol, or expression denoting the power to which the symbol or number is to be raised or multiplied by itself.

exponential decay (Ionizing Radiation)   A mathematical expression describing the rate at which a radioactive material decays.

exposed (Epidemiology)   Those who have been exposed to a supposed cause of a disease or possess a characteristic that is a determinant of the health outcome of interest.

exposure   The proximity to a condition that may produce injury or damage. Contact with a chemical, physical, or biologic hazard that may or may not have the potential to cause an adverse health effect from inhalation, skin contact, or other route of entry of the reagent into the body.

exposure (OSHA)   Condition that occurs when an employee is subjected to a toxic substance or harmful physical agent in the course of employment through any route of entry, and that includes past exposure and potential exposure, but does not include situations in which the employer can demonstrate that the toxic substance or harmful physical agent is not present, used, handled, stored, or generated in the workplace in any manner different from typical non-occupational situations.

exposure assessment   A determination, either qualitatively or quantitatively, of the magnitude of the hazard potential of a task or job, based on the frequency and duration of exposure, the properties of the materials handled/used of health concern (e.g., vapor pressure, temperature, toxicity), process factors, (e.g., open/closed system), exposure control methods in effect, and other factors that may affect the magnitude of the exposure and the dose of the contaminant received.

exposure limit   The concentration of an airborne contaminant to which a worker can be exposed for a specified time limit, such as the 8-hour work day or for 15 minutes, without adverse effect.

exposure rate   The exposure per unit time, such as parts per million-hours (ppm-hr).

exposure route   The manner b which a material enters the body, such as by inhalation, ingestion, etc.

external radiation   Exposure to ionizing radiation from a source outside the body.

external respiration   The exchange of oxygen and carbon dioxide between the lung air and blood.

Extraction Procedure Toxicity Test   [EP]   Toxicity test for determining the toxicity of a waste relative to EPA's CERCLA requirements.

extraneous   Irrelevant, superfluous, and not essential.

extrapolation   A calculation, based on limited data from natural or experimental observation of humans or other organisms exposed to a substance, that aims to estimate the dose-effect relationship outside the range of the available data. The estimation of a mathematical function at a point which is larger than all the points at which the value of the function is known.

extremely low frequency magnetic field   A magnetic field with a frequency in the range of 0 to 3000 hertz that results from current flowing in electrical conductors.

extremities   The hands and forearms, and the feet, ankles and legs below the knees.

extrinsic   Existing external to an individual and not forming an essential part of it.

eye protection   Devices that protect the eyes in an eye-hazard environment. The use of safety glasses, splash goggles, or other protective eyewear that will reduce the potential for eye contact with a hazardous material being used/handled/processed.

eyewash fountain   A device used to irrigate and flush the eyes in the event of eye contact with a hazardous substance. Water flow rate and period of use are recommended for emergency situations.

F

f Feet.

f Femto, 1 E $^{-15}$.

f Fibers.

F Degrees Fahrenheit.

fabric collector (Ventilation) Air cleaning device designed to remove particulates from exhaust air by straining, impingement, interception, diffusion, or electrostatic charge.

fabric filter A filter media for removal of particulates from an air stream passed through it.

face (Ventilation) The opening in an exhaust hood or enclosure through which the exhaust air enters.

facepiece That part of a respirator which covers the wearer's nose, mouth, and, in a full facepiece, the eyes.

face shield A protective device designed to prevent hazardous materials, dusts, sharp objects, and other materials from contacting the face. A device worn in front of the eyes and a portion of, or all of, the face. It supplements the eye protection afforded by a primary protective device (e.g., safety glasses).

face velocity Velocity of the air at the plane of the opening into a hood or other planned entry (slot, canopy, etc.) into a ventilation system.

facial hair policy A policy that does not permit the presence of facial hair that could prevent a good respirator face seal on personnel who may be required to wear respiratory protection. Respirators are not to be worn when conditions prevent a good face seal. Such conditions may include the presence of a beard, long sideburns, a mustache, or other facial hair growth. Some facilities do not permit such facial hair on anyone who comes on the site.

facies The front aspect of the head.

facility (OSHA) The buildings, containers, or equipment that contain a process.

FACOSH Federal Advisory Council for Occupational Safety and Health.

F.A.C.P. Fellow of the American College of Physicians.

F.A.C.S. Fellow of the American College of Surgeons.

factor Agent, condition, or environmental variable that contributes to or modifies an action or effect.

Factory Mutual Association [FM] An industrial fire protection, engineering, and inspection bureau established and maintained by mutual insurance companies. The Factory Mutual laboratories test and list fire protection equipment for approval, assist in the development of standards, and conduct research in fire protection.

facultative saprophytes Organisms that can only survive on dead organic matter.

Fahrenheit temperature scale [F] The scale of temperature in which 212° is the boiling point of water at 760 millimeters of mercury pressure, and 32° is the freezing point.

fail safe   Design of a device or system in such a manner that if it fails or ceases to operate, it will do so in a safe position or condition.

faint   To have a temporary loss of consciousness as a result of a reduced supply of blood to the brain. Also referred to as syncope.

fallout (General)   Particulate materials from the atmosphere that settle on fixed surfaces in the environment.

fallout (Ionizing Radiation)   Radioactive debris from a nuclear detonation that becomes airborne or has deposited on the earth. The dust and other particulate material that contains radioactive fission products from a nuclear explosion.

fall time (Instrument)   The time interval between an initial response in an instrument and a specified percent decrease (e.g., 90 percent) after a decrease in the inlet concentration.

false negative   A result that is positive, but has been recorded as negative in a data set.

false positive   A result that is negative, but has been recorded as positive in a data set.

FAM   Fibrous aerosol monitor.

familial   Occurring in or affecting more members of a family than would be expected by chance.

fan   A mechanical device that creates static pressure in order to move air through it.

fan, airfoil   A type of backward-inclined blade fan with blades that have an airfoil cross-section.

fan, axial   A fan in which airflow is parallel to the fan shaft and air movement is induced by a screwlike action of the fan blade.

fan, backward-inclined blade   A centrifugal fan with blades inclined opposite to fan rotation.

fan, centrifugal   A fan in which the air leaves the fan in a direction perpendicular to the direction of entry.

fan curve   A curve relating pressure produced by a fan to the flow rate (cfm) of a given fan at a fixed speed.

fan, forward-curved blade   A centrifugal fan with blades inclined in the direction of fan rotation.

fan laws   Statements and equations that describe the relationship between fan volume, pressure, brake horsepower, size, and rotating speed for application when changes are made in fan operation. For example, volume varies directly as fan speed, TP, and SP vary as the square of fan speed, and horsepower varies as the cube of the fan speed.

fan, paddle wheel   A centrifugal fan with radial blades.

fan, propeller   An axial fan employing a propeller to move air.

fan, radial blade   A centrifugal fan with radial blades extending out radially from the fan wheel shaft.

fan rating table   Tables published by fan manufacturers presenting the range of capacities of particular fan models along with the static pressure developed and the fan speeds within the limits of the fans' construction.

fan, squirrel cage   A centrifugal blower with forward-curved blades.

fan static pressure   The static pressure to be added to that of the ventilation system due to the presence of the fan. It equals the sum of pressure losses in the system minus the velocity pressure in the air at the fan inlet.

fan, tube axial   An axial fan mounted in a duct section.

fan, vane axial   An axial flow fan mounted in a duct section with vanes to straighten the airflow and increase static pressure.

F.A.P.H.A.   Fellow of the American Public Health Association.

far field (Acoustics)   The uniform sound field that is free and undisturbed by bounding surfaces and other sources of sound and in which the sound pressure level obeys the inverse-square law relationship and decreases 6 dB for each doubling of distance from the source. Also referred to as a free sound field.

farmer's lung   A dust disease, or pneumoconiosis, resulting from the inhalation of moldy silage.

farsightedness   The ability to see objects better from a distance than from short range.

fasciculation   A small local contraction of muscles, visible through the skin, representing a spontaneous discharge of a number of fibers innervated by a single motor nerve.

fatality rate   The death rate observed in a designated group of persons who have been affected by a simultaneous event.

fathom   A unit of length equal to 6 feet, used principally in the measurement and specification of marine depths.

fatigue (Metal)   The tendency of a metal to break under repeated loading at a level below that expected based on a static test.

fatigue (Physiological)   Physical or mental weariness or exhaustion resulting from exertion.

fatty acid   A carboxylic   (-COOH)   acid derived from or contained in an animal or vegetable fat or oil that may be either solid, semisolid, or liquid.

fault tree analysis   A method for the analysis of hazards or potentially hazardous situations using the fault tree analysis approach (i.e., considering all possible factors that may contribute to the occurrence of an undesired event in sequence, diagraming these in the form of a tree with branches continued until all the probabilities of the undesired event and the chain of events leading up to them are determined).

fc   Foot-candle.

f/cc   Fibers per cubic centimeter of air.

FDA   U.S. Food and Drug Administration.

feasibility study   A gathering of information to determine if a proposed project is practical and capable of being completed successfully.

feasible   A measure that is practical and capable of being accomplished or brought about.

febrile   Pertaining to or characterized by fever.

fecal   Of, pertaining to, or constituting feces.

fecal coliform bacteria   Organisms associated with the intestines of warm-blooded animals. Their presence indicates contamination by fecal material and the potential presence of organisms capable of causing human disease.

feces Waste excreted from the bowels.

fecundity The physiological ability to reproduce.

Federal Emergency Management Agency [FEMA] An independent agency that advises the president on meeting civil emergencies and provides assistance, when recommended by the president, to individuals and public entities that suffered property damage in emergencies and disasters.

Federal Housing Administration [FHA] Federal agency whose function has been transferred to the Department of Housing and Urban Development.

Federal Insecticide, Fungicide, and Rodenticide Act [FIFRA] An act established by the EPA to regulate the registration, production, storage, use, transportation, and disposal of these pesticides.

Federal Register [FR] A daily publication of the U.S. government in which official documents, promulgated under the laws, are published. In addition, advanced notice of proposed rule makings are published here, along with notices of meetings/hearings related to these.

feedlot A concentrated, confined poultry or animal growing operation in which the animals or poultry are fed at the place of confinement.

FEF Forced expiratory flow.

FEMA Federal Emergency Management Agency.

femto [f] Prefix indicating 1 E^{-15}.

Feret's diameter The distance between the extreme boundaries of a particle.

fermentation A chemical change, usually the decomposition of sugars and starches, that is induced by living organisms or enzymes occurring in unicellular plants, such as yeasts, molds, and fungi.

ferruginous bodies *See* asbestos bodies.

fertile material A material that is not fissile itself, but can be converted into a fissile material by irradiation in a reactor.

fertility toxin A substance which reduces male or female fertility.

fertilizer A substance or mixture that contains one or more of the primary nutrients nitrogen, phosphorous, and potassium.

fetal Of or pertaining to the fetus.

fetal death rate The number of fetal deaths per 1000 births over a stated time period.

fetid Having an offensive odor.

fetotoxic Any agent that is toxic to the fetus.

fetotoxicity The property or ability of a substance to produce a toxic effect to a fetus.

fetotoxin A substance that is toxic to the fetus.

fetus The unborn human from the end of the eighth week of pregnancy to the moment of birth.

FEV Forced expiratory volume.

FEV$_1$ Forced expiratory volume–one second.

fever A condition in which the body temperature is above normal.

FHA   Federal Housing Administration.

fiber (General)   A particle having a length to diameter/width ratio of greater than 3 : 1, and a length of 5 micrometers or more.

fiber (PCM Method)   Particulate at least 5 micrometers in length with an aspect ratio (length to width ratio) of at least 3 : 1. A rodlike structure having a length of at least three times its diameter.

fiber (EPA–TEM Method)   Structure greater than or equal to 0.5 micrometers in length with an aspect ratio of 5 : 1 or greater, and having substantially parallel sides.

fiber optics   A system of flexible quartz or glass fibers with internal reflective surfaces that can transmit light.

fibrillation   Rapid and uncoordinated contractions of the heart.

fibrils   Small slender fibers.

fibroblast   Connective tissue cell.

fibrosis   The formation and accumulation of fibrous tissue, especially in the lungs. Also the chronic collagenous degeneration of the pulmonary parenchyma.

fibrosis-producing dust   A dust that, when inhaled, deposited, and retained in the lungs, can produce fibrotic growth that may result in pulmonary disease.

fibrotic   The abnormal formation of fibrous tissue.

fibrous   Description of a material which contains fibers.

fibrous aerosol monitor   [FAM]   A direct-reading instrument for measuring the concentration of airborne fibers.

FID   Flame ionization detector.

field blank sample   Sampling media, such as a charcoal tube, filter cassette, or other device, that is handled in the field in the same manner as are other sampling media of the same type, but through which no air is sampled. These are used in sampling and analysis procedures to determine the contribution to the analytical result from the media plus any contamination that may have occurred during handling in the field, shipping, and storage before analysis. Often referred to as a blank sample.

field duplicate sample   A sample that is collected concurrently with another sample of the same type, and in the same location for the same duration.

FIFRA   Federal Insecticide, Fungicide, and Rodenticide Act.

Filar micrometer   A microscopic attachment used for determining the size of particles.

film badge   A packet of photographic film that is used to measure exposure to ionizing radiation. The badge may contain two or three types of film of differing sensitivities, as well as filters to shield the film for determining exposure to various types of radiation.

film ring   A film badge in the form of a finger ring that is typically worn by personnel whose hands may be exposed to ionizing radiation during use of a radiation source (e.g., operation of an X-ray diffraction unit).

filter (Acoustics)   A device for separating components of a signal on the basis of their frequency. Electronic circuitry designed to pass or reject a specific frequency band.

filter (Respirator)   The media component of a respirator that removes particulate materials, such as dusts, fumes, fibers, and/or mists from inspired air.

filter (Sampling)   Sampling media for collection of airborne particulate contaminants in order to determine the concentration of the material in the air. Filter media may be made of cellulose fibers, glass fibers, mixed cellulose esters (membrane filter), polyvinyl chloride, Teflon, polystyrene, silver, or other material.

filter (Ventilation)   Media, such as plastic, paper, or cloth, through which air is passed for the removal of suspended particulate matter.

filter aid   Powdery substances, such as diatomaceous earth, that are used to improve the effectiveness of a filter media.

filter efficiency   The efficiency of a filter media expressed as collection efficiency (percentage of total particles collected) or as penetration (percent of particles that pass through the filter).

filtering facepiece   Disposable dust mask. A negative-pressure particulate respirator with a filter as an integral part of the facepiece or with the entire facepiece composed of the filtering medium.

filter media (Sampling)   The material, such as PVC, glass-fiber, molecular membrane, etc., that is used for collecting airborne particulate samples.

filtrate   The liquid that is passed through a filter.

filtration (Respiratory Protection)   The process of removing a contaminant from air being inhaled.

filtration (Sampling)   The process of collecting a contaminant on an appropriate filter media for determining its composition and concentration in the sampled air, as well as for determining if the exposure level is acceptable or whether exposure controls must be developed and implemented.

fines   Particulates of small size. Typically used in conjunction with small aerosols.

fire   The process of rapid oxidation that generally produces both heat and light. Also referred to as combustion.

fire area   An area of a building that is separated from the remainder of the building by construction, having a fire resistance of at least one hour and having all communicating openings properly protected by an assembly having a fire rating of at least one hour.

fire brigade   An organized group of employees who are knowledgeable, trained, and skilled in at least basic firefighting operations.

fire classes   See fires.

firedamp   A combustible gas, usually methane, occurring naturally in coal mines and forming explosive mixtures with air.

fire damper   Device installed in ductwork in which the duct passes through fire separations to aid in preventing the spread of a fire.

fire door   Doors that building codes and the NFPA classify according to a rating system related to their fire resistance, such as 3-hour, 1 1/2-hour, 1-hour, etc.

fire extinguishers, classes (Portable Hand Held)   Class A extinguishers rated for Class A fires may contain water, aqueous-film-forming foam, dry chemical, or a halogen compound;

Class B extinguishers for Class B fires may contain carbon dioxide, dry chemical, aqueous-film-forming foam, or halogenated agents; Class C extinguishers for Class C fires may contain carbon dioxide, dry chemical, or a halogenated compound; and Class D extinguishers for use on Class D fires contain specially prepared materials, such as Met-L-X powder, T.E.C. powder, etc.

fire gases   Gases that remain when products of combustion are cooled to normal temperature.

fire partition   A partition that serves to restrict the spread of a fire, but does not qualify as a fire wall.

fire point   The minimum temperature to which a material must be heated to sustain combustion after ignition by an external source.

fire resistance rating   The time period, in hours or fractions thereof, that a building member or assembly will resist a code-specified fire test without failure. For example, a 2-hour rating indicates that the assembly withstood the standard test for greater than two hours without failure by any of the failure criteria listed in the fire test protocol.

fire resistive   The quality of a structure or material to provide a predetermined degree of fire resistance, usually rated in hours.

fire retardant   Chemicals, paints, or coatings used for the treatment of combustible building materials that provide a lesser degree of protection than a fire-resistant material would have.

fires   Classes of fires: Class A fires are fires in ordinary combustible materials (e.g., paper, wood, cloth); Class B fires are those in combustible or flammable liquids, flammable gases, greases, and similar materials; Class C fires are those in electrical equipment; and Class D fires are those in combustible metals (e.g., magnesium, sodium, potassium, etc.).

fire separation   The distance in feet measured from the building face to the closest interior lot line, to the center of a street or public way, or to an imaginary line between two buildings on the same property.

firestopping   The barriers for restricting the spread of fire in concealed spaces; the materials to fill gaps around penetrations in walls and ceilings.

fire triangle   The three elements that must be present in the right proportion for a fire to exist: oxygen (or an oxidizing agent), fuel (or a reducing agent), and heat. Keeping the three elements of the fire triangle apart is the key to preventing fires, and removing one or more of these elements is the key to extinguishing fires that do start.

fire wall   A fire-resistance-rated wall that is erected to restrict the spread of fire.

fire watch   Individual provided with fire extinguishing equipment and training in its use who is posted (1) in the immediate area of welding or cutting operations when other than a minor fire might develop; (2) where there is an appreciable amount of combustible building material nearby; (3) where sparks could ignite nearby combustibles; or (4) where combustible materials are adjacent to the opposite side of metal partitions, walls, ceilings, or roofs and are likely to be ignited by conduction or radiation.

first aid (OSHA)   Any one-time treatment and any follow-up visit for the purpose of observation of minor scratches, cuts, burns, splinters, and so forth that do not ordinarily require medical care.

first aid injury   An injury requiring first aid treatment only, such as when only one treatment and only one subsequent observation of the injury is required.

fiscal year   [FY]   A 12-month period for which an organization plans the use of its funds.

fissile material   Any material fissionable by thermal neutrons.

fission   The splitting of a nucleus into at least two other nuclei with the release of a relatively large amount of energy.

fissionable material   A material that can be fissioned (split) into other nuclei by any process.

fission gases   Fission products, such as radon, krypton, and xenon, that exist in the gaseous state.

fission products   The products produced as a result of the fissioning of heavy elements, plus the nuclides formed by the fission fragments' radioactive decay.

fit check (Respiratory Protection)   Action by a respirator user to determine if the respirator is properly seated to the face. A positive or negative pressure check is made to ensure proper respirator facepiece seal. *See* user seal check.

fit factor (Respiratory Protection)   A measure of how well a respirator fits as determined by a quantitative fit test. The ratio of the test agent concentration outside the respirator to the agent concentration inside the respirator.

fit test (Respiratory Protection)   A protocol for determining an individual's ability to obtain a good fit with a particular size and style of respirator. Fit testing may be qualitative or quantitative.

fixed contamination (Radioactivity)   Contamination that is unlikely to become airborne so that it may be inhaled or unlikely to adhere to the hands on being touched, so that it is not likely to be ingested.

flame arrester   Device used in gas vent lines and other similar locations to arrest or prevent the passage of flame into an enclosed space, such as a container or flammable liquid storage cabinet.

flame ionization detector   A carbon detector that relies on the detection of ions formed when a carbon-containing material, such as a volatile or gaseous hydrocarbon, is burned in a hydrogen-rich flame. This detector is commonly used in a gas chromatograph to detect and quantitate organic compounds. It is also employed in some portable instruments.

flame photometric detector   A detection system based on the luminescent emissions between 300 and 425 nanometers when sulfur compounds are introduced into a hydrogen-rich flame. An optical filter system is used to differentiate the sulfur compounds present from other materials. This detector finds application in gas chromatography.

flameproofing material   Chemicals that catalytically control the decomposition of cellulose material at flaming temperature.

flame propagation   The spread of a flame throughout an entire volume of a vapor-air mixture from a single source of ignition.

flame retardant   The use of chemicals, paints, or coatings for the treatment of materials to retard both the rate of burning and the rate at which fuel is contributed by the treated material.

flame spread   The propagation of flame over a surface.

flame spread rating   The surface burning characteristics of various materials as described in NFPA 101.

flammable   Any substance that is easily ignited and burns, or has a rapid rate of flame spread. Capable of being ignited and of burning. Substance with a flash point below 100°F.

flammable limits   The percent by volume limits (i.e., upper and lower flammable limits) of a flammable gas/vapor at normal temperature and pressure in air above and below which flame propagation does not occur on contact with a source of ignition. See flammable range.

flammable liquid   Liquid having a flash point below 100°F and a vapor pressure not exceeding 40 pounds per square inch (absolute) at 100°F is a Class I flammable liquid. Class I flammable liquids are subdivided into Class IA, which have a flash point below 73°F and a boiling point below 100°F; Class IB, which have a flash point below 73°F and a boiling point above 100°F; and Class IC, which have a flash point at or above 73°F and below 100°F.

flammable range   All concentrations of a mixture of a flammable vapor or gas in air in which a flash will occur or a flame will travel if the mixture is ignited. The lowest percentage at which this will occur is the lower explosive limit, and the highest percentage is the upper explosive limit. The difference between the lower and upper flammable limits, expressed in terms of percentage of a vapor or gas in air or oxygen by volume, is the range. *See* flammable limits.

flammable solid   A solid material that is easily ignited and that burns rapidly.

flammable storage cabinet   A cabinet for the storage of flammable and combustible liquids that is constructed in accordance with the requirements for storage cabinets for flammable and combustible liquids in NFPA 30.

flange   A blank plate extending around a hood or part of it to minimize the amount of air that enters the exhaust hood from behind or on the sides, thereby maximizing that which enters from the front.

flare   Typically, a stack with a pilot light at the top that is employed for the release of combustible/flammable vapors/gases as a flame at an elevation above ground. Its use is normally associated with an upset condition in a process operation. In some locations ground flares are employed. Elevated flares are often a community noise concern.

flash-back arrester   A device utilized on a vent for a flammable liquid or gas storage container to prevent flashback into the container when a flammable or explosive mixture ignites outside the container.

flash blindness   Temporary visual disturbance resulting from viewing an intense light source.

flash burn   Inflammation of the cornea as a result of exposure to the ultraviolet light associated with welding operations, particularly electric-arc welding, as well as exposure to direct and reflected radiation from ultraviolet lamps.

flashover   The stage of a fire at which all surfaces and objects are heated to their ignition temperature and flame breaks out almost at once over the entire surface.

flash point   [fl.p.]   The lowest temperature at which a liquid gives off enough vapor to form an ignitable mixture with air and to produce a flame when a source of ignition is present. Two methods, referred to as the open cup and closed cup methods, are available for determining the flash point of a material.

flexion   Movement in which the angle between two bones connecting to a common joint is reduced.

flexor muscles   Muscles that, when contracted, decrease the angle between limb segments.

flocculant   A substance that induces flocculation, or the aggregation of suspended colloidal particles, in such a way that they form small clumps.

flocculate   *See* agglomeration.

floppy disk   A flexible plastic disk that is a common form of external data storage for a microcomputer system.

flora   The plant life present in or characteristic of a specific location.

flow contour (Ventilation)   Lines of equal air velocity in front of a hood opening.

flow diagram   A diagram consisting of blocks connected by line arrows that represent the series of steps and the direction of flow in a process. *See* block-flow diagram.

flow meter   Device for measuring the amount of fluid (air, gas, or liquid) flowing through it.

flow rate   The volume per time unit (e.g., liters per minute, etc.) given to the flow of air or other fluid by the action of a pump, fan, etc.

fl.p.   Flash point.

flue   A pipe or other channel through which combustion air, smoke, steam, or other material is vented to the atmosphere.

flue gas   The emissions from a combustion process that are typically discharged from a stack or flue.

flue gas scrubber   A pollution control device that is used to remove fly ash and other contaminants that are generated in a combustion process or other operation and are otherwise released to the atmosphere through a stack or flue.

fluence   Flowing smoothly and easily.

fluid   A substance that flows and yields to any force tending to change its shape.

fluidization   A technique in which finely divided solids are caused to behave like a fluid by suspending them in a moving gas or liquid. An example is the fluid catalytic cracking process in petroleum refineries.

fluorescence   The emission of electromagnetic radiation, especially that of visible light, as a result of the absorption of electromagnetic radiation, and persisting only as long as the stimulating radiation is continued.

fluorescent screen   A screen coated with a fluorescent substance so that it emits light when irradiated with X-rays.

fluoridation   The addition of fluoride to the public water supply as a public health measure for preventing tooth decay.

fluoroscope   A device consisting of a screen covered with crystals of calcium tungstate on which is projected the shadows of X-rays passing through a body that is placed between the screen and the source of irradiation.

fluorosis   Disease characterized by mottling of the teeth, resulting from exposure to airborne fluoride dusts.

flux (Electromagnetic Radiation)   The radiant or luminous power of a light beam.

flux (Ionizing Radiation)   The amount of some type of radiation passing through a specific area per unit time. The unit of flux is the number of particles, energy, etc. per square centimeter per second.

flux (Soldering)   A substance used to clean the surface and promote fusion in a soldering procedure, and to prevent the formation of oxides.

flux density (Ionizing Radiation)   The number of photons passing through one square centimeter of a surface per second.

fly ash   Particulate entrained in flue gas that is emitted as a result of fuel combustion, particularly in which coal is burned.

f/m   Feet per minute.

FM   Factory Mutual Engineering Association.

foam   A frothlike mixture used in fire fighting.

focal length (Microscope)   The distance from the optical center of the lens to its focus.

fog   Condensed water vapor in cloudlike masses close to the ground that limits visibility.

follicle   An approximately spherical group of cells containing a cavity.

folliculitis   The inflammation of follicles, particularly hair follicles.

follow-up (Toxicology)   The time of observation from the first and/or last exposure in an experimental study.

follow-up study (Epidemiology)   A study in which the individuals or populations selected on the basis of whether they had (1) been exposed to a risk, (2) received a specified preventive or therapeutic procedure, or (3) possessed a certain characteristic, are followed to determine the outcome of the exposure, the procedure, or the effect of the occurrence of a disease.

fomites   Intimate personal articles, such as clothing, a drinking glass, a handkerchief, etc., that can be contaminated and transmit disease agents.

font   The size and style of type.

Food and Drug Administration   An agency of the Department of Health and Human Services that oversees the safety of food, cosmetics, and drugs to ensure they are safe, pure, and wholesome, and made under sanitary conditions.

foodborne disease   Diseases resulting from the consumption of food in which specific disease-causing microorganisms have grown.

food handler   Individual who comes into contact with food or utensils used in the manufacture, storage, preparation, handling, or serving of food.

food poisoning   Acute illness resulting from the consumption of contaminated food. Contamination may be from toxins in the food or the result of the presence of bacteria.

food vendor   Any person who sells, offers for sale, or gives away from any vehicle, container, or market any article of food for human consumption.

foot-candle   [fc]   Unit of illumination. The quantity of light on a plane surface of one foot from and perpendicular to a standard candle. Multiply foot-candles by 10 to get the approximate corresponding metric unit value in lux.

foot-lambert   English unit of illuminance. Multiply foot-lamberts by 35 to get the approximate candles per square meter value.

foot-pound   [ft-lb]   A unit of work equal to the work done by a force of 1 pound acting through a distance of 1 foot in the direction of the force.

force   The capacity to do work or cause physical change.

forced draft   The positive pressure created by air being blown into a furnace or other combustion equipment by a fan or blower.

forced expiratory volume–one second   [FEV_1]   The maximum volume of air that can be forced from an individual's fully inflated lungs in 1 second.

forced vital capacity   [FVC]   The volume of air that can be forcibly expelled from the lungs after a full inspiration of air.

forecasting   A method of estimating what may happen in the future based on reliance on the extrapolation of existing information and trends.

foreseeable emergency   Any potential occurrence, such as equipment failure, failure of a control system, a release of a toxic material as a result of a seal failure, etc., that could result in an uncontrolled release of a hazardous chemical.

Fortran   A high-level computer language designed for scientific and mathematical use, with the name Formula Translator and the acronym Fortran.

fossil fuel   Fuel, such as natural gas, petroleum, coal, etc., that originated from the remains of plant, animal, and sea life of previous geological eras.

fp   Freezing point.

FPD   Flame photometric detector.

fpm   Feet per minute.

fps   Feet per second.

FR   Federal Register.

free area (Ventilation)   The total area of the open space in a grill or diffuser through which air can pass.

free fall velocity   The final equilibrium velocity of a particle falling freely through a fluid.

free radical   A molecular fragment having one or more unpaired electrons, usually short-lived and highly reactive.

free silica   Silica in the form of cristobalite, tridymite, or alpha quartz.

free sound field   *See* far field.

freeze protected deluge shower   A deluge shower that is designed to operate at temperatures that would freeze water in the system.

freeze trap (Sampling)   A method to collect gases/vapors by cooling the sampled air to a temperature at which the substance(s) of interest condense and are thus collected.

freezing point   [fp]   The temperature at which the crystals of a material are in equilibrium with its liquid phase at atmospheric pressure.

frequency   The number of cycles, revolutions, or vibrations completed per unit of time. Typically indicated as cycles per second (cps) or revolutions per minute (rpm). The rate at which oscillations are produced.

frequency (Acoustics)   The number of cycles occurring per second of a periodic sound or vibration, designated as hertz.

frequency analyzer (Acoustics)   Instrument having frequency selecting/filtering capability for analyzing the spectral content of sound.

frequency distribution   The tabulation of data from the lowest to the highest, or highest to the lowest, along with the number of times each of the values was observed or occurred in the distribution.

frequency of exposure   The number of times per shift, day, year, etc. that an individual is exposed to a harmful substance or physical agent.

frequency rate (Disabling Injury)   Relates the injuries that occur to the hours worked during the period and expresses them in terms of a million man-hour unit.

friable asbestos   An asbestos-containing material (i.e., material that contains more than 1 percent asbestos) that can be crumbled, pulverized, or reduced to powder by hand pressure when dry.

friable material   Any material used in construction that, when dry, can be crumbled or reduced to a powdery consistency by hand pressure.

friction   The resistance to movement, such as the rubbing of one object or surface against another.

friction loss (Ventilation)   The pressure loss in a ventilation system due to friction of the moving air on the ductwork.

frit   The porous section at the end of a glass tube which is employed in a glass flask to break up an air stream into small bubbles, thereby improving the absorption of air contaminants by the sorbent as it is sampled through it. Often referred to as a glass frit.

frit (Ceramics)   Finely ground inorganic minerals mixed with fluxes and coloring agents that turn into a glass or enamel on heating.

fritted bubbler   *See* glass frit.

front   The boundary at which air masses of different temperatures and densities collide.

frostbite   The freezing of tissue with consequent disruption of cell structure.

FRP   Fiberglass-reinforced plastic.

ft   Foot or feet.

ft²   Square foot or square feet.

ft³   Cubic foot or cubic feet.

ft/min   Feet per minute.

ft/s   Feet per second.

fuel cell   A device for converting chemical energy into electrical energy.

fugitive dust   Particulate matter composed of soil uncontaminated by pollutants that results from industrial activity. Typically considered to be any airborne particulate matter emitted to the atmosphere other than from a stack.

fugitive emissions   The release of airborne contaminants into the surrounding air other than through a stack, chimney, vent, or other opening, including release from sources such as the sealing mechanisms of pumps, compressors, flanges, valves, and other type seals. Thus, fugitive emissions result from an equipment leak and are characterized by a diffuse release of materials such as VOCs, hydrocarbons, etc. into the atmosphere. The EPA defines fugitive emissions as those emissions that do not occur as part of the normal operation of the plant.

full facepiece respirator   A respirator that covers the wearer's entire face from the hairline to the chin.

full scale (Instrument)   The maximum measurement value or maximum limit for a given range on an instrument.

full shift   The regularly scheduled work period, typically of 8 hours duration.

fume cupboard   The British term for a laboratory fume hood.

fume fever   *See* metal fume fever.

fume hood   The type of hood employed in chemical laboratories with a sliding safety-glass sash in the front that can be opened when working within the hood. Typically is provided with its own exhaust fan for discharging exhaust air above the lab roof.

fumes   Small, solid particles formed by the condensation of the vapors of solid materials.

fumigant   An agent for exterminating vermin or insects.

fumigation   The use of a fumigant to destroy pests.

functional group (Chemistry)   A relatively reactive group of atoms that is responsible for the characteristic reactions of a compound. May include the hydroxyl group, carboxyl group, the amide group, and many others.

fundamental frequency (Acoustics)   The lowest periodic frequency component present in a complex spectrum.

fungi   A group of lower plants that lack chlorophyll and live on dead and living organisms. Includes molds, mildews, mushrooms, etc.

fungicide   A substance that destroys or inhibits the growth of fungi.

fungistat   Any agent that inhibits the growth of fungi.

furuncle   A painful nodule formed in the skin by inflammation of the dermis and subcutaneous tissue as a result of staphylococci bacteria which enter through hair follicles. Also referred to as a boil.

fusible plug   A safety relief device in the form of a plug of low-melting metal that is intended to yield or melt at a set temperature to permit the release of the material in the line/vessel/etc.

fusion   The act of coalescing or joining two or more atomic nuclei. A nuclear reaction characterized by the joining together of light nuclei to form heavier nuclei, the energy for the reactions being provided by the agitation of particles at high temperatures.

FVC   Forced vital capacity.

FVC–1   Forced vital capacity–one second.

G

g Gram(s).

G Gauss.

G Giga, 1 E9.

Gaia hypothesis An hypothesis that states that the oxygen/carbon dioxide balance established and maintained by organisms determines and maintains the earth's climate and other environmental conditions.

gage pressure The pressure with respect to atmospheric pressure or above atmospheric pressure as indicated on an appropriate pressure gage.

gain (Instrument) The ratio of the signal output to input. Gain is frequently referred to as span.

gal Gallon(s).

galvanic cell An electrolytic cell whose electric effect is brought about by the difference in electrical potential between two dissimilar metals.

galvanize The depositing of a protective coating of zinc on a metal by dipping it in molten zinc.

gamete A germ cell possessing the haploid number of chromosomes.

gamma ray Short wavelength electromagnetic radiation of nuclear origin. Gamma rays are highly penetrating and present an external radiation hazard.

ganglion A knot or knotlike mass.

gangrene Death and decay of tissue in a part of the body, usually a limb, due to failure of blood supply, injury, or disease.

gas A state of matter in which the material has a very low density and viscosity, can expand and contract greatly in response to changes in temperature and pressure, easily diffuses into other gases, and readily and uniformly distributes itself throughout any container. Formless fluid that tends to occupy an entire space uniformly at ordinary temperatures and pressures.

gas chromatography An analytical chemical procedure involving passing a sample through a column of specific makeup to separate the components of the sample, enabling them to elute or pass out of the column separately and be detected and quantified by one or more detectors, such as a flame ionization detector, thermal conductivity detector, electron capture detector, etc.

gas chromatograph–mass spectrometer Refers to both an analytical method as well as the apparatus used in the analysis. The gas chromatograph serves to separate the components of the sample, and the mass spectrometer serves to identify them by exposing the eluted components to a beam of electrons, which causes ionization to occur. The ions produced are accelerated by an electric impulse, passed through a magnetic field, separated, and identified based on their mass.

gas constant The constant R in the perfect gas equation ($PV = RT$).

gas free A tank, compartment, or other type containment or area that has been tested, using appropriate instruments, and found to be sufficiently free, at the time of the test, of toxic or explosive gases/vapors for a specified purpose.

gas frit   A sintered or fritted glass surface that is designed to break up an air stream into small bubbles in order to increase the contact of the air with a liquid sorbent, thereby improving the absorption of specific gaseous contaminants present in the air. *See* also frit.

gasification   The conversion of a solid or liquid fuel into a gaseous fuel.

gas laser   A type of laser in which the laser action takes place in a gas medium, such as carbon dioxide.

gas mask   A face-covering, respiratory protective unit which is provided with its own air purifying device for removing specific harmful contaminants from the inspired air.

gasohol   A mixture of gasoline and ethyl alcohol that is used as an auto fuel.

gasoline   A blend of light hydrocarbon fractions of relatively high antiknock value, with proper volatility, clean-burning characteristics, and containing additives to prevent rust and oxidation, and with sufficiently high octane rating to prevent knocking. Gasolines typically contain some benzene.

gas pressure   The force exerted by a gas on its surroundings.

gas test   An analysis of the air to detect unsafe concentrations of toxic or explosive gases/vapors.

gastric   Pertaining to, affecting, or originating in the stomach.

gastric lavage   Washing out the stomach with a tube and fluids.

gastritis   Chronic or acute inflammation of the stomach.

gastroenteritis   Inflammation of the mucous membrane of the stomach and intestines.

gastrointestinal tract   [GI tract]   The system consisting of the stomach, intestines, and related organs.

gas/vapor detection instrument   An assembly of electrical, mechanical, and often chemical components that senses and responds to the presence of a gas/vapor in air mixture.

gate (Guard)   A movable barrier arranged to enclose the point of operation on a machine before the stroke of a press can be initiated.

gauge pressure   The difference between two absolute pressures, such as the pressure above atmospheric. *See* gage pressure.

gauss   [G]   The centimeter–gram–second electromagnetic unit of magnetic flux density, equal to 1 maxwell per square centimeter, or 1 tesla equal to 1 E^4 gauss.

Gaussian curve   A normal statistical curve with even distribution on either side of the mean.

gavage (Toxicology)   Dosing an animal by introducing a test material through a tube into the stomach.

gBq   Gigabecquerel, 1 E^9 Bq.

GC   Gas chromatograph or gas chromatography.

GC–ECD   Gas chromatography–electron capture detector.

GC–FID   Gas chromatography–flame ionization detector.

GC–FPD   Gas chromatography–flame photometric detector.

GC–MS   Gas chromatograph–mass spectrometer as an instrument, or gas chromatography – mass spectrometry as a method.

GC–PID   Gas chromatography–photoionization detector.

GC–TCD   Gas chromatography–thermal conductivity detector.

Geiger counter   *See* Geiger–Mueller counter.

Geiger–Mueller counter   [G–M counter/tube]   A highly sensitive, gas-filled radiation-measuring device. More commonly referred to as a Geiger counter or Geiger tube.

gene   Fundamental unit of inheritance which determines and controls hereditarily transmissible characteristics.

general duty clause (OSHA)   A clause that states that employers are to furnish employees a place of employment that is free from hazards that are causing or are likely to cause death or serious physical harm.

general duty clause violation (OSHA)   A violation that exists when OSHA can show that the hazard was a recognized hazard, that the employer failed to render its workplace free from the recognized hazard, that the occurrence of an accident or adverse health effect was reasonably foreseeable, that the likely consequence of the incident (accident or adverse effect) was death or a form of serious physical harm, and that there existed feasible means to correct the hazard.

general environment (Ionizing Radiation)   The total terrestrial, atmospheric, and aquatic environment outside sites within which any activity, operation, or process authorized by a general or special license is performed.

general exhaust ventilation   A mechanical system for exhausting of air from a work area, thereby reducing the contaminant concentration by dilution.

general license (Ionizing Radiation)   A license issued by the NRC or an Agreement State for the possession and use of certain radioactive materials, often for small quantities, for which a specific license is not required. Individuals are automatically licensed when they buy or obtain a radioactive material from a vendor who has a license from the NRC to sell products containing small amounts of some radioactive materials.

general ventilation   A system of natural or mechanically induced fresh air movement designed to mix with and dilute contaminants in the workroom air. General ventilation is used typically for the control of temperature, humidity, or odors, but not to control exposure to toxic substances. Synonomous with dilution ventilation.

generation rate   The amount of pollutant generated over a stated time period.

generic formula (Chemistry)   Expresses a generalized type of organic compound in which the variables stand for the number of atoms or the kind of radical in a homologous series, such as ROR for an ether, ROH for an alcohol, or C_nH_{2n} for an olefin.

generic name   A nonproprietary name, such as the chemical identity of a material or product rather than identification by a registered trade name.

genesis   The coming into being of anything.

genetic   Pertaining to or involving genes.

genetic code   The information code that determines the nature and type of organism, i.e., its heredity, and the kind of cell structure to be formed.

genetic defect   A defect in a living organism resulting from a deficiency in the genes of the original reproductive cells from which the organism was conceived.

genetic effects   Inheritable changes, chiefly mutations, produced by the absorption of ionizing radiation, exposure to certain chemicals, ingestion of some medications, and other causes.

genetic engineering   The manipulation of the array of the genes of a living organism.

genetics   The biology of heredity, especially the study of mechanisms of hereditary transmission and variation of organismal characteristics.

genome   The complete set of hereditary factors, as contained in the haploid assortment of chromosomes.

genotoxicity   The ability of a toxicant to interact with DNA.

genotoxin   A substance that is toxic to genetic material.

genotype   All or part of the genetic constituents of an organism.

genus   A taxonomic classification below a family and above a species; to form the name of the species.

geochemistry   The study of the chemical composition of the earth in terms of the physicochemical and geological processes and principles that produce and modify minerals and rocks.

geologist   Scientist who gathers and interprets data pertaining to the strata of the earth's crust.

geology   Science related to the study of the structure, origin, history, and development of the earth and its inhabitants as revealed from the study of rocks, formations, and fossils.

geometric mean   The median in a lognormal distribution, arrived at by taking the sum of the logarithms of the individual values, calculating their arithmetic mean, and converting it back by taking the antilogarithm of the arithmetic mean.

geophysics   The study of the earth's physical properties, including the air and space.

geosphere   The solid earth, which supports most plant life.

geothermal energy   Superheated water and steam trapped in rock strata in areas characterized by volcanic activity or by intrusions of molten magma.

g-eq   Gram equivalent.

geriatrics   The branch of medicine that treats all problems peculiar to old age and the aging.

germ (Medicine)   A microorganism, especially a pathogen.

germ cell   The cell of an organism whose function it is to reproduce the kind (i.e., an ovum or spermatozoon). The cells of an organism whose function is reproduction.

germicide   An agent that kills pathogenic organisms.

gerontology   The scientific study of the physiological and pathological phenomena associated with aging.

gestation period   The period of carrying developing offspring in the uterus after conception.

GFCI   Ground fault circuit interrupter.

GFF   Glass fiber filter.

GFI   Ground fault interrupter.

GHG   Greenhouse gases.

GHz   Gigahertz, 1 E^9 Hz.

G.I.   Gastrointestinal.

giddiness   A dizzy or light-headed sensation.

giga   [G]   Prefix indicating 1 E^9.

gingivitis   Inflammation involving the gums.

gland   An organ that extracts specific substances from the blood and concentrates or alters them for subsequent secretion.

glanders   A contagious disease of horses that is transmissible to man.

glare   Brightness within the field of vision that causes annoyance, discomfort, eye fatigue, and/or interference with vision. Any extraneous scattering of light.

glass frit   *See* frit.

glassification   A means of disposing of some hazardous wastes by mixing them with components of glass, then heating the mixture to make a glass which immobilizes the waste and resists leaching.

glaucoma   A disease of the eye characterized by high intraocular pressure, damaged optic disk, hardening of the eyeball, and partial, or complete loss of vision, caused by aging, infection, disease, or congenital defect. A leading cause of blindness.

GLC   Ground level concentration.

globe thermometer   A thermometer with its bulb positioned in the center of a painted (flat black) metal sphere. Used to measure radiant heat temperature in assessing heat stress.

glossy   Describing a polished surface with a mirrorlike finish.

glovebag   A plastic bag that is placed around a pipe or other structure from which the removal of a material, such as asbestos, is to be carried out without its release to the atmosphere.

glove box   A sealed enclosure in which materials are handled through long impervious gloves fitted to openings in the walls of the enclosure. There are several classes of glove boxes for various applications and necessary contamination control.

GLP   Good laboratory practice.

glycogen   The chief carbohydrate that is formed by and largely stored in the liver, and, to a lesser extent, in muscles. It is depolymerized to glucose and liberated as needed by the body.

G–M counter   Geiger–Mueller counter.

goggles   A device with contoured eyecups or with full facial contact, having glass or plastic lenses, and held in place by a headband or other suitable means to provide protection of the eyes and eye sockets.

goniometer   An apparatus for measuring the limits of flexion (bending) and extension of the joints of the fingers.

good work practices   Practices that include using good judgment, following established guidelines and procedures, exercising care appropriate to the risk, and adhering to safety.

gore (Ventilation)   Triangular section of sheet metal that is used to make elbows for ductwork.

gpm   Gallons per minute.

gr   grain(s).

grains per cubic foot   gr/cu. ft.

grab sample   An air sample collected over a short period of time (e.g., minutes), providing an indication of a contaminant concentration at a specific time.

Grade D breathing air   Breathing air which meets the specifications of the Compressed Gas Association Comodity Specification for Grade D air. It must have between 19.5 and 23 percent oxygen, not more than 5 milligrams per cubic meter of condensed hydrocarbons, no more than 10 parts per million of carbon monoxide, a maximum of 1000 parts per million of carbon dioxide, and no pronounced odor.

grain   [gr]   A unit of weight equal to 65 milligrams.

gram   [g]   A metric unit of weight.

gram–atomic weight   [gram–atom]   The atomic weight of an element expressed in grams.

gram–calorie   *See* calorie.

gram–equivalent weight   [g–eq]   The equivalent weight expressed in grams. *See* equivalent weight.

gram–mole   *See* gram–molecular weight.

gram–molecular weight   [gram–mole]   The molecular weight of a compound in grams.

granulocytes   Any cell containing granules, especially a leukocyte containing neutrophil, basophil, or eosinophil granules in its cytoplasm.

granulocytosis   An abnormally large number of granulocytes in the blood.

granuloma   A tumorlike mass or nodule of vascular tissue due to a chronic inflammation process associated with an infectious disease.

graph   A visual display of the relationship between variables.

graticule   *See* reticle.

gravimetric   Pertaining to measurement by weight.

gravimetric method   An analytical method for determining the concentration of a substance based on the determination of the weight of the material collected on a filter, absorbed in a sorbent, or formed in a subsequent analytical procedure.

gravity separator (Ventilation)   A chamber or housing in which the velocity of the exhaust air is made to drop rapidly so that dust particles settle out by gravity.

gray   [Gy]   A unit of absorbed radiation dose equal to 1 joule per kilogram.

graywater   Sink, bath, and shower water, excluding that from the toilet.

Greenburg–Smith impinger   A large impinger that has been employed for the collection of airborne dust samples.

grille   A covering over an air inlet or outlet with openings through which air passes.

grinder's asthma   Asthmatic symptoms related to the inhalation of fine particles generated in the grinding of metals.

grinder's consumption   The term for silicosis that occurs among workers using silica-based grinding wheels.

ground   A conductive connection, whether intentional or accidental, by which an electric circuit or equipment is connected to reference ground.

ground fault circuit interrupter   [GFCI]   A protective electrical device that senses electric current leakage from a ground fault and immediately breaks the affected circuit. Intended for the protection of personnel by de-energizing the circuit or part of it when the current to ground exceeds a predetermined value.

ground fault interrupter   *See* ground fault circuit interrupter.

grounding   The practice of eliminating the difference in voltage potential between an object and ground. Procedure involves connecting the object to an effective ground (metal to metal) by an appropriate wire.

ground level concentration   [GLC]   The concentration of airborne contaminants at about 4–5 feet above grade.

ground state   The lowest energy level of an atom.

groundwater   The supply of fresh water under the earth's surface that forms a natural reservoir of this resource.

GSA   General Services Administration.

guard   A barrier that prevents entry of an operator's hands or fingers into the point of operation of equipment or a machine.

guarded   Covered, shielded, fenced, enclosed, or otherwise protected to remove the likelihood of approach or contact by persons to an existing or potential danger.

guarding   Any means of effectively preventing personnel from coming in contact with the moving parts of machinery or equipment which could cause physical harm to them.

guardrail   A rail secured to uprights and erected along the exposed side and ends of a platform.

guillotine damper   A heavy plate, installed vertically in a breeching, to regulate the flow of gases.

Gy   Gray.

h   Hecto, 1 E^2.

h   Hour(s).

Haber's Rule   A rule that states that toxic effect is dependent upon the product of exposure time and the contaminant concentration. Thus, exposure at a higher concentration for a short period would be equivalent to exposure at a lower concentration for a longer period in direct proportion to the product of exposure concentration and time. This reportedly holds true, however, only for short exposure periods. Also referred to as Haber's Law.

habitat   The area or type of environment in which an organism or biological population normally lives or occurs.

hair cells   Sensory receptors responsible for the initiation of neural impulses in the auditory nerve.

hail   Precipitation in the form of frozen rain drops.

half-life (Biological)   *See* biological half-life.

half-life (General)   Time in which the concentration of a substance is reduced 50 percent.

half-life (Radioactive)   Time required for a radioactive substance to lose one-half of its activity by radioactive decay. Each radionuclide has a unique half-life.

half-life (Toxicology)   The time required for one-half of the dose of a substance to be excreted from the body.

half-mask respirator   Respirator which covers half the face, from the bridge of the nose to below the chin.

half thickness (Radiation)   The thickness of a shielding material necessary to reduce the intensity of a beam of ionizing radiation to one-half its initial value. Also referred to as the half-value layer.

half-value layer (Radiation)   [HVL]   The thickness of a specified material which, when introduced into the path of a given beam of ionizing radiation, reduces the exposure rate by one-half. Also referred to as the half-thickness.

hallucination   A perception of an external stimulus object in the absence of such an object.

hallucinogen   A substance which induces hallucinations.

hallucinogenic   Capable of producing hallucinations.

halocarbon   Carbon compounds with one or more halogen atoms as a part of the substance. These are believed to deplete the concentration of ozone in the stratosphere.

halogen   Any of the elements fluorine, chlorine, bromine, and iodine.

halogenated extinguishing agents   Hydrocarbons in which one or more hydrogen atoms has been replaced by atoms of the halogen series. This substitution confers nonflammability as well as flame extinguishing properties, which makes these agents useful in portable fire extinguishers and in fixed extinguishing systems.

halons   Term given to a group of hydrocarbon compounds in which one or more hydrogen atoms is replaced by atoms of the halogen series of elements. Typically used as fire extinguishing agents.

hand-held drench shower   A flexible hose connected to a water supply and used to irrigate and flush eyes, face, and body areas in the event of contact with a hazardous material that is corrosive, irritating, absorbed through the skin, etc.

handicap (General)   A reduction of an individual's ability to fulfill a social role as the result of an impairment, inadequate training for the role, or other circumstance.

handicap (Hearing)   Any hearing impairment that is sufficient to affect a person's efficiency in the activities of daily living.

hand protection   Gloves or other type hand protection that prevent the harmful exposure of the wearer to hazardous materials.

haploid   Having the number of chromosomes present in the normal germ cell equal to half the number in the normal somatic cell.

HAP(S)   Hazardous air pollutant(s).

hardware   The physical equipment that makes up a computer system.

hard water   Water that contains dissolved minerals, does not lather with soap easily, and forms insoluble deposits in boilers and heat exchangers.

hardwired   A system in which there is a direct connection of components by wires or cables.

harmful   Term indicating the potential for an agent or condition to produce injury or an adverse health effect.

harmonic (Acoustics)   A tone in the harmonic series of overtones that is produced by the fundamental tone. A frequency component at a frequency that is an integer multiple of the fundamental frequency.

HAVS   Hand arm vibration syndrome.

hazard (Industrial Hygiene)   A hazard is a condition with the potential for causing injury or sickness to personnel, damage to equipment or structures, loss of material, or lessening of the ability of personnel to perform a function, but without consideration of the consequence. Hazard is the estimated potential of a chemical or physical agent, ergonomic stress, or biologic organism to cause harm based on the likelihood of exposure, the magnitude of exposure, and the toxicity or effect. A material poses a hazard if it is likely that an individual will encounter a harmful exposure to it.

hazard (Safety)   A dangerous condition, potential or inherent, that can interrupt or interfere with the expected orderly progress of an activity.

hazard and operability study   [HAZOPS]   A formal, structured, investigative system for examining potential deviations of operations from design conditions that could create process operating problems and hazards.

hazard awareness   Information and training provided by employers to employees on the hazardous chemicals in their work areas at the time of initial assignment and whenever a new chemical is introduced in their work assignments or work areas.

hazard class   A group of materials, as designated by the Department of Transportation, that share a common major hazardous property, such as flammability, radioactivity, etc.

hazard classification   Designation of relative accident potential based on the likelihood that an accident will occur.

hazard communication (OSHA)   Regulation to ensure that the hazards of all chemicals produced or imported are evaluated and that information concerning their hazards is transmitted to employers and employees via a comprehensive hazard communication program.

hazardous   Dangerous or perilous.

hazardous air pollutant   [HAP]   An air pollutant to which no ambient air quality standard is applicable and which, in the judgment of the administrator, causes or contributes to air pollution that may reasonably be expected to result in mortality or an increase in serious irreversible or incapacitating reversible illness.

hazardous atmosphere   Any atmosphere that is oxygen deficient or contains toxic or other type health hazards at concentrations exceeding established exposure limits. Also considered to be an atmosphere that may expose personnel to the risk of death, incapacitation, or the impairment of their ability for self-rescue, injury, or illness.

hazardous chemical   A chemical for which there is statistically significant evidence, based on at least one study conducted in accordance with established scientific principles, that acute or chronic health effects may occur in exposed employees. Also considered to be any chemical which is caustic, explosive, flammable, poisonous, corrosive, reactive, or radioactive, and which requires special care when handling because its presence or use is a physical or health hazard.

hazardous chemical (OSHA)   Any chemical that is a physical or chemical hazard.

hazardous condition   Circumstances that are causally related to an exposure to a hazardous material.

hazardous decomposition   The breaking down or separation of a substance into its constituent parts or simpler compounds, accompanied by the release of heat, gas, or hazardous material.

hazardous locations (National Electric Code)   Areas that require specially designed electrical equipment because of the presence or potential presence of flammable liquids or gases, combustible dusts, or readily ignitable fibers or flyings. Such locations are divided into three classes depending upon the kind of hazardous material involved. Each class is divided into divisions and subdivisions according to the likelihood of a potential hazard existing and the specific substance. The type of electrical equipment, such as lighting fixtures, electric switches, circuit breakers, etc. for use in the hazardous location is specified for each class.

hazardous material   Any substance or compound that has the ability to produce an adverse health effect in a worker or a safety problem.

Hazardous Material Identification System   [HMIS]   A comprehensive communication program involving hazard assessment, labeling, material safety data sheets, and employee training through the use of colors, numbers, letters, and symbols.

hazardous waste   Solid or liquid waste that exhibits any of the following characteristics: ignitability, corrosivity, reactivity, or EP (extraction procedure) toxicity. A more general definition is a waste that may cause or contribute to mortality or may pose a threat to human health.

Hazardous Waste Operations and Emergency Response (OSHA)   [HAZWOPER] Regulation covering various types of hazardous waste cleanup operations where there is a reasonable likelihood that employees will be exposed to safety and health hazards.

hazard ratings   A method for conveying information on the health effects of a chemical, its flammability potential and reactivity properties, and for conveying special warnings regarding carcinogenicity, radioactivity, what not to use in case of fire, etc. The NFPA diamond symbol is such a method for providing this type information.

hazard ratio   The airborne concentration of particulates divided by the exposure limit.

hazard ratio (NIOSH)   The ratio of the air concentration of a contaminant to its PEL.

hazard recognition   The perception of a hazardous condition, such as is made in an exposure assessment of a job or task.

hazard warning   The means employed to convey information on the hazardous properties of a material, such as presenting the information, including symbols, words, pictures, or combinations of these or other forms of warning, on a label or other appropriate medium, to convey the specific physical or health hazard, including the target organ.

hazardous waste site   Any facility or location at which hazardous waste operations take place.

haze   A decrease in visibility due to moisture, dust, smoke, and vapors suspended in the air to form a partially opaque condition.

HAZMAT   Acronym for hazardous material.

HAZMAT Team (OSHA)   An organized group of employees, designated by the employer, who are expected to perform work to control actual or potential leaks or spills of hazardous substances requiring possible close approach to the substance.

HAZOPS   Hazard and operability study.

HAZWOPER   Hazardous Waste Operations and Emergency Response (an OSHA regulation).

Hb   Hemoglobin.

HbCO   Carboxyhemoglobin.

HbO$_2$ Oxyhemoglobin.

HBV   Hepatitis B virus.

HC   Hydrocarbon(s).

HCP   Hearing conservation program.

head   Term used for indicating pressure, such as a head of 1-inch water gauge.

health (WHO)   A state of complete physical, mental, and social well-being; not merely the absence of disease or infirmity.

health education   A procedure by which individuals and groups learn to behave in a manner-conducive to the promotion, maintenance, and restoration of health.

health hazard   A property of a chemical, mixture of chemicals, physical stress, pathogen, or ergonomic factor for which there is statistically significant evidence, based on at least one test or study conducted in accordance with established scientific principles, that acute or chronic adverse health effects may occur among workers exposed to the agent. Health hazards include chemicals which are carcinogens, toxics, irritants, corrosives, reproductive toxins, hepatotoxins, nephrotoxins, neurotoxins, skin sensitizers, and hemolytic agents; and physical agents such as noise, heat stress, non-ionizing radiation, and vibration; as well as biologic agents and ergonomic factors.

health physicist   [HP]   An individual trained in radiation physics, its associated health hazards, the means to control exposures to this physical hazard, and the establishment of procedures for work in radiation areas.

health physics   The discipline devoted to the protection of humans and the environment from unwarranted exposure to ionizing radiation and non-ionizing radiation.

Health Physics Society   [HPS]   Professional society of persons active in the field of health physics, a profession devoted to the protection of people and their environment from radiation hazards.

health professional (OSHA)   A physician, occupational health nurse, industrial hygienist, toxicologist, or epidemiologist providing medical or other occupational health services to exposed employees.

healthy worker effect   A phenomenon observed in studies of occupational diseases in which workers exhibit lower death rates than the general population because the infirmed, severely ill, and many disabled have been excluded from employment, and because those who are employed are generally healthy.

hearing   The subjective response to sound, including the entire mechanism of the external, middle, and inner ear, and the nervous and cerebral operations which translate the physical operations into meaningful signals.

hearing conservation   Measures taken to prevent or minimize the loss of hearing among employees. A hearing conservation program involves assessing employee noise exposure risk, providing exposed personnel hazard awareness training, installing engineering controls on significant noise-producing sources, implementing administrative controls if necessary, providing noise-exposed personnel appropriate hearing protection when necessary to reduce their exposure to noise by this means, and instituting a hearing test program for determining the effectiveness of the overall program.

hearing conservation program   *See* hearing conservation.

hearing impaired   A person with a hearing loss sufficient to affect efficiency in the course of everyday living.

hearing level   The deviation in decibels of an individual's hearing threshold at various test frequencies as determined by an audiometric test based on an accepted standard reference level.

hearing loss   The difference in decibels of an individual's hearing threshold at various test frequencies from the zero reference of the audiometer. The types of hearing loss are conductive, sensory and neural. Exposure to noise produces, primarily, sensory hearing loss.

hearing protection   *See* hearing protective device.

hearing protective device   Any device or material that is capable of being worn and that reduces the amount of sound energy entering the outer ear and proceeding to the receptors in the inner ear.

hearing threshold   The weakest or minimally perceived sound, in decibels, that an individual can detect during an audiometric test at a particular time.

hearing threshold shift   The change in the intensity level at which sound can be heard. The shift can be either a permanent threshold shift (PTS) or a temporary threshold shift (TTS). The standard threshold shift is considered to be a change in hearing threshold relative to a baseline audiogram of an average of 10 decibels or more at 2000, 3000, and 4000 hertz in either ear.

heat   A form of energy that raises the temperature of an object or person.

heat balance   A condition in which the heat lost from the body is equal to the heat gain. Keeping such a balance is essential to maintain homeostasis.

heat cramps   *See* cramps.

heat exchanger   A device that is used to transfer heat from one medium to another.

heat exhaustion   Circulatory failure in which the venous blood supply that is returned to the heart is significantly reduced, and, as a consequence, fainting may result. Early symptoms can include fatigue, headache, dizziness, nausea, shortness of breath, high pulse rate, and irritability.

heat gun (Paint)   An apparatus that emits heat with enough intensity to soften paint sufficiently to enable it to be scraped from a surface.

heating, ventilating, and air conditioning system   [HVAC system]   The system that is in place to provide ventilation, heating, cooling, dehumidification, humidification, control of odors, and cleaning of the air for maintaining comfort, safety, and health of the occupants of a building, workspace, etc.

heat of combustion   The difference between the heats of formation of the starting fuel and that of the combustion products.

heat of reaction (Chemistry)   The quantity of heat released in a chemical reaction.

heat prostration   *See* heat exhaustion.

heat pump   A mechanical refrigerating system used for cooling in the summer and which, when the evaporator and condenser effects are reversed, absorbs heat from the air or water in winter, thereby supplying heat.

heat rash   Inflammation of the sweat glands, which have become plugged due to the skin swelling, thereby impairing sweating and diminishing a worker's ability to tolerate heat.

heat sink   A device or substance that is used to absorb or dissipate unwanted thermal energy.

heat strain   A physical condition that occurs when heat balance is not maintained and the heat stress on the body becomes excessive. Heat strain can then result, with consequent heat cramps, heat exhaustion, or heat stroke.

heat stress   The physiologic effect on the body that can result from exposure to excessive heat. Excessive heat stress may lead to heat stroke, heat exhaustion, heat cramps, or prickly heat depending on the temperature, humidity, conditioning of the person exposed, ventilation, rest breaks, temperature of the rest area, availability of tempered water, and other factors.

heat stress index   [HSI]   An engineering approach for determining the stress on individuals working in hot environments. It is based on the ratio of the amount of heat loss that can occur from sweating (to maintain thermal equilibrium or homeostasis) to the maximum evaporative capacity of the work environment. The maximum evaporative capacity of the environment is determined from the moisture content of the air, air movement, and dry-bulb temperature at the work location, as well as on the clothing the individual is wearing.

heat stroke   One of the consequences of exposure to heat-stressful situations in which the body temperature rises to a high level (e.g., over 105°F), the thermoregulatory function fails, and sweating stops. The body temperature can rise to a critical level, and death may result. Heat stroke is the most severe form of the heat stress disorders. Prompt medical intervention is essential for the heat stroke victim.

heat study   The determination of the heat stress on workers exposed to a hot environment by using one of the heat stress evaluation methods, such as the Wet-Bulb Globe Temperature procedure or the Heat Stress Index method. Measurements typically carried out at the worker location include determinations of the dry-bulb temperature, wet-bulb temperature, natural wet-bulb temperature, radiant heat temperature, relative humidity, and air movement.

heat wave   A weather condition that exists if there are three consecutive days with a temperature reading of 90°F or above.

HEG   Homogeneous exposure group.

helmet (Respiratory Protection)   A rigid respiratory inlet covering that also provides head protection against impact and penetration.

helminth   A worm, especially one that is parasitic.

hematemesis   Vomiting of blood.

hematocrit   The percent by volume of erythrocytes in whole blood.

hematologic toxin   Toxic substance that affects the blood or blood-forming organs.

hematologist   An individual trained in the science encompassing the generation, anatomy, physiology, pathology, and therapeutics of blood.

hematology   The branch of medical science concerned with the generation, anatomy, physiology, and therapeutics of blood. The study of the form and structure of blood and blood-forming organs.

hematoma   A localized collection of blood, usually clotted, in an organ, space, or tissue, due to a break in the wall of a blood vessel.

hematopoetic   Pertaining to or affecting the formation of blood cells.

hematopoetic changes   Changes in the formation of blood cells.

hematopoetic system   The blood-forming system in the body.

hematotoxicity   The toxic effects of various substances and physical agents on blood and blood-forming organs.

hematuria   The presence of blood in the urine.

hemialgia   Pain affecting one side of the body only.

hemoglobin   [Hb]   The oxygen-carrying pigment of erythrocytes that is formed by the developing erythrocyte in bone marrow.

hemoglobinuria   The presence of free hemoglobin in the urine.

hemolysis   The breakdown of red blood cells with the liberation of hemoglobin.

hemolytic   Pertaining to, characterized by, or producing hemolysis.

hemolytic anemia   Anemia resulting from the excessive destruction of red blood cells.

hemoptysis   Expectoration of blood or of blood-stained sputum.

hemorrhage   Profuse loss of blood from blood vessels and/or capillaries.

hemotoxin   A poison that causes the lysis, or destruction, of red blood cells.

henry   [H]   A unit of conductance in which an induced electromotive force of 1 volt is produced when the current is varied at the rate of 1 ampere per second.

Henry's law   A physical law that states that when a liquid and gas remain in contact, the weight of gas that dissolves in a given quantity of the liquid is proportional to the pressure of the gas above the liquid.

HEPA   High-efficiency particulate air (filter).

hepatitis   Inflammation of the liver.

hepatitis B   A disease that can lead to cirrhosis, liver cancer, and often death.

hepatitis B virus   A virus that causes an acute infection of the liver, resulting in jaundice and symptoms referable to liver damage. Early symtoms of infection by hepatitis B virus include fever, nausea, fatigue, headache, and abdominal discomfort.

hepatomegaly   Enlargement of the liver.

hepatotoxin   Substance that causes damage to the liver, such as carbon tetrachloride.

herbicide   A substance used to kill plants, especially weeds.

heredity   Transmission of characteristics and traits from parent to offspring.

hernia   The protrusion of a loop or knuckle of an organ or tissue through an abdominal opening.

hertz   [Hz]   A unit of frequency equal to 1 cycle per second.

heterotrophic bacteria   Bacteria that depend upon a reduced form of organic compounds for both energy and the carbon required to build their biomass.

HEW   U.S. Department of Health, Education, and Welfare.

HFE   Human factors engineering.

HHC   Highly hazardous chemical.

HiEF   High-efficiency particulate filter.

high altitude   An elevation over 4000 feet.

high-boiling aromatic oils   [HBAO]   These are high-boiling components produced during catalytic cracking and thermal cracking of petroleum streams, and also during the extraction of lube base stocks. They contain complex mixtures of hydrocarbons in the boiling range of 500 to 1000°F and have demonstrated carcinogenic potential in animal testing. These are also referred to as aromatic process oils.

high-efficiency particulate air filter   [HEPA filter]   A filter that is at least 99.97 percent efficient in removing an aerosol of monodisperse dioctylphthalate with a diameter of 0.3 micrometers.

high frequency loss (Acoustics)   Hearing loss in frequency bands of 2000 hertz and above. Also referred to as high frequency hearing loss.

highly hazardous chemical (OSHA)   [HHC]   Chemical listed in Appendix A of the OSHA standard related to process safety management of highly hazardous chemicals (i.e., 29 CFR 1910.119). A substance possessing toxic, reactive, flammable, or explosive properties.

high radiation area   An area in which personnel could receive a dose equivalent of greater than 100 millirems per hour at 30 centimeters from a source or surface that the radiation penetrates.

high-volume air sampler   [hi-vol]   Sampling device used for the collection of particulates in the ambient air. One type is employed for collecting PM 10 particulates (i.e., those equal to or less than 10 micrometers in diameter), and another is for collecting all suspended particulates in order to determine the total suspended particulate concentration.

histamine   An amine found in the body that has three functions: dilation of capillaries, which increases capillary permeability and reduces blood pressure; constriction of the bronchial smooth muscles of the lungs; and induction of increased gastric secretions.

histogram   A graphic presentation of the frequency distribution of a variable by drawing rectangles in such a manner that their bases lie on a linear scale representing different intervals, and their heights are proportional to the frequencies of the values within each of the intervals of the base.

histology   The study of the structure of tissues.

histopathology   Pathologic histology, or the change in the function of tissues as a result of a disease.

histoplasmosis   Bacterial infection resulting from the inhalation of the spores of *Histoplasma capsulatum*. Occupations at risk are those associated with the raising and processing of fowl.

HIV   Human immunodeficiency virus.

hives   *See* urticaria.

hi-vol   *See* high-volume air sampler.

HMIS   Hazardous Material Identification System of the National Paint and Coatings Association.

holistic   Viewing, treating, and emphasizing the importance of the whole and the interdependence of its parts.

hologram   A three-dimensional image produced on a photographic plate employing a laser system.

homeostasis   A state of physiologic equilibrium within the body.

homogeneous   Uniform in construction or composition throughout.

homogeneous exposure group   [HEG]   A group of employees who experience exposures similar enough so that monitoring the exposure of any of the group will provide exposure data that is useful for predicting the exposures of the remainder of the group.

homologous series   A series of organic compounds in which each successive member has one more CH_2 group in its molecule than the preceding member.

HON   Hazardous Organic NESHAP.

hood (Respiratory Protection)   A respiratory inlet covering that completely covers the head and neck and may also cover portions of the shoulders and torso.

hood (Ventilation)   A shaped inlet designed to capture contaminated air and conduct it into a duct of an exhaust ventilation system.

hood, canopy   A hood located over a source of emission.

hood, capturing   A hood with sufficient airflow to reach outside the hood and draw in contaminants.

hood, enclosing   A hood that encloses the source of contamination.

hood entry loss   The pressure loss from turbulence and friction as air enters a ventilation system hood.

hood, receiving   A hood that is sized and positioned to catch a stream of contaminants or contaminated air directed at the hood.

hood, slot   A hood provided with a narrow slot(s) leading into a plenum chamber that is under suction to distribute the intake of air along the length of the slot(s).

hood static pressure   The energy necessary to accelerate air from rest outside a hood to duct velocity inside the duct, as well as the losses associated with the air entering the hood (i.e., hood entry losses). It represents the suction that is available to draw air into the hood.

horsepower [hp]   A unit of measure of work done by a machine equal to 745.7 watts or 33,000 foot-pounds per minute.

hose mask   Respiratory protective device that supplies air to the wearer from an uncontaminated source through a hose that is connected to the facepiece.

host   One that harbors an organism that is not ordinarily parasitic in the particular species and from whom it derives its food supply and shelter.

host factors   The personal characteristics of individuals who harbor or nourish a parasite.

hot (Ionizing Radiation)   Term describing an area, piece of equipment, or other item as being highly radioactive.

hot spot (Ionizing Radiation)   An area or location where the surface contamination is noticeably greater than in neighboring regions in the area.

hot-tap   Any connection in a pipeline or other equipment that is either under pressure or has not been cleared and prepared for tapping.

hot work   Mechanical or other work that involves a source of heat, sparks, or other source of ignition that is sufficient to cause ignition of a flammable material. Work involving sources of ignition or temperatures high enough to cause the ignition of a flammable mixture. Examples include welding, burning, soldering, using power tools, operating engines, and sandblasting; the operation of electric hot plates, explosives, open fires, and portable electrical equipment that has not been tested and classified as intrinsically safe; and other sources of ignition.

hot-work permit   A document issued by an authorized person permitting specific work for a specified time, to be done in a defined area employing tools and equipment that could cause the ignition of a flammable substance.

housekeeping   The maintenance of the orderliness and cleanliness of an area or facility.

hp   Horsepower.

HP   Health physics.

HPD   Hearing protective device.

HPLC   High performance liquid chromatography.

HPS   Health Physics Society.

hr   Hour(s).

HR   Hazard ratio.

HSI   Heat Stress Index.

HUD   U.S. Department of Housing and Urban Development.

human carcinogen   Substance that has been shown by valid, statistically significant epidemiological evidence to be carcinogenic to humans.

human factors engineering   [HFE]   The science having to do with man, his work, and the efficient use of human beings. A subdiscipline of ergonomics that is concerned with the design of procedures, equipment, and the work environment to minimize the likelihood of an accident/disease to occur.

human immunodeficiency virus   [HIV]   A retrovirus which compromises the ability of the body to deal with various infections and forms of cancer by damaging the autoimmune system. The virus causes acquired immunodeficiency syndrome.

humidification   The process of adding moisture to an airstream.

humidifier   A device for adding water vapor to the air.

humidifier fever   Respiratory disease resulting from exposure to organisms dispersed into the air from a humidifier.

humidity   The water vapor content of the air.

humidity, absolute   The weight of water vapor per unit volume of air, such as pounds per cubic foot or grams per cubic centimeter.

humidity, relative   The ratio of the actual partial pressure of the water vapor in a space to the saturation pressure of pure water at the same temperature.

humidity, specific   The weight of water vapor per unit weight of dry air.

humoral   Pertaining to or arising from any of the four body fluids: blood, phlegm, choler, and black bile.

HVAC   Heating, ventilating, and air conditioning.

HVAC system   Heating, ventilating, and air conditioning system.

HVL   Half-value layer.

HWM   Hazardous waste material.

hydrocarbon   Any of a large class of organic compounds containing only carbon and hydrogen.

hydrogenation   A reaction in which hydrogen is added to a substance by the use of gaseous hydrogen. The process is accomplished by means of a catalyst and proceeds more rapidly at high pressure.

hydrologic cycle   The movement of water between the atmosphere, earth, and oceanic systems through the processes of precipitation, evaporation, infiltration, and stream flow.

hydrology   The science dealing with the properties, distribution, and flow of water on or in the earth.

hydrolysis   A chemical reaction in which water reacts with another substance to form two or more new substances, such as the formation of a weak acid or base or both.

hydrometer   An instrument used for determining the specific gravity of liquids.

hydrophilic   Having an affinity for, absorbing, tending to combine with, or capable of dissolving in water.

hydrophobic   Substances that are incapable of dissolving in water.

hydrophytic   Plant that grows in water or is adapted to a very wet environment.

hydroponics   A technique for growing plants through the use of nutrient solutions and without use of soil.

hydrosphere   The waters of the earth as distinguished from the lithosphere and the atmosphere.

hydrostatic   Relating to fluids not in motion, especially incompressible fluids and the pressure they exert or transmit.

hydrostatic testing   A method for testing the integrity of a cylinder that contains a substance under pressure.

hygrometer   An instrument used to measure the water vapor content of the air.

hygrometry   The determination of the water vapor content of the air.

hygroscopic   A substance that has the property of absorbing moisture from the air.

hygrothermograph   A recording instrument which provides a simultaneous reading of ambient temperature and humidity.

hyperbaric   Air pressure in excess of that at sea level.

hypercapnia   An excessive amount of carbon dioxide in the blood.

hyperemia   An excess of blood in tissue, organ, or other part of the body.

hyperglycemia   Abnormally increased content of sugar in the blood.

hypergolic   Liquid rocket fuel or propellant consisting of a combination of fuel and oxidizer which ignite spontaneously on contact with each other.

hyperhidrosis   Excessive perspiration.

hyperhydration   A state of excessive water content of the body.

hyperkeratosis   Hypertrophy of the horny layer of the skin.

hyperpigmentation   Abnormally increased pigmentation or skin darkening that occurs among some people as a result of exposure to photosensitizers such as coal tar and other chemicals, some medications, and ultraviolet light.

hyperplasia   The abnormal multiplication or increase in the number of normal cells in normal arrangement in tissue.

hypersensitivity   A state of abnormal responsiveness (stronger than normal) of the body to a foreign agent or chemical.

hypersensitivity pneumonitis   A complex inflammatory response resulting from exposure to organic materials, commonly fungi or thermophilic bacteria. The response is focused in the respiratory bronchioles and alveoli with fibrosis as a possible end result. It is not an infection or an allergenic response, thus is somewhat of a misnomer.

hypersusceptibility   An unusually high response to a dose of a substance.

hypertension   Persistently high arterial blood pressure.

hyperthermia   Abnormally high body temperature which may lead to the development of heat stress effects, such as heat exhaustion, heat cramps, or heat stroke.

hypertrophy   The enlargement or overgrowth of an organ or part due to an increase in size of its constituent cells.

hyperventilation   Abnormally prolonged, rapid, and deep breathing. This results in reduced carbon dioxide in the blood (acapnia) and consequent apnea (intermittent cessation of breathing).

hypnotic   Substance that induces sleep or sleepiness.

hypobaric   Air pressure below that at sea level.

hypoglycemia   An abnormally diminished content of glucose in the blood, which may lead to tremulousness, cold sweat, hypothermia, and headache, accompanied by confusion, hallucinations, bizarre behavior, and, ultimately, convulsions and coma.

hypokinetic hypoxia   A shortage of oxygen due to interference with blood flow.

hyponatremia   A deficiency of sodium in the blood.

hypophrenia   Mental retardation.

hypopigmentation   A loss of pigment in the skin.

hypoplasia   The incomplete development of an organ so that it fails to reach adult size.

hyposensitive   Abnormally decreased sensitivity, such as that which is developed by repeated and gradually increasing doses of an offending material.

hyposusceptibility   An unusual low response to some dose of a substance.

hypotension   Abnormally low blood pressure. Typically seen in shock, but not necessarily indicative of it.

hypotensive agent   A stimulus or substance that produces hypotension.

hypothermia   Reduced body temperature. An acute effect due to prolonged exposure to cold with resultant rapid loss of heat from the body.

hypothesis   An assumption that may be accepted or rejected based on experimental findings, such as by statistical tests of significance.

hypoxemia   Deficient oxygenation of the blood.

hypoxia   A state of inadequate oxygenation of the blood and tissues. Anemic hypoxia is the reduction of the oxygen-carrying capacity of the blood as a result of a decrease in the total hemoglobin or as the result of an alteration of the hemoglobin constituents.

hysteresis   The maximum difference in output for any given input when the value is approached first with increasing input signal, then with decreasing input signal. The non-uniqueness in the relationship between two variables as a parameter increases or decreases.

Hz   Hertz.

I

IAEA   International Atomic Energy Agency.

IAP   Indoor air pollutant/pollution.

IAQ   Indoor air quality.

IARC   International Agency for Research on Cancer.

iatrogenic   The creation of additional problems or complications resulting from treatment by a physician or surgeon.

IBP   Initial boiling point.

IC   Integrated circuit.

IC   Ion chromatography.

ICC   U.S. Interstate Commerce Commission.

ICRP   International Commission on Radiological Protection.

icterus   Jaundice due to the deposition of bile pigment in the skin and mucous membranes with a resulting yellow appearance of the individual.

i.d.   Inside diameter.

ID   Intradermal.

idiomatic   Peculiar to or characteristic of a given language.

idiopathic   A disease of unknown origin or cause.

idiosyncrasy   A habit or quality of body or mind peculiar to any individual.

IDLH   Immediately dangerous to life or health.

IEEE   Institute of Electrical and Electronics Engineers.

IES   Illuminating Engineering Society.

ignitable   Capable of being set on fire.

ignitable waste   A waste that poses a fire hazard during routine storage, handling, or disposal.

ignitability characteristics   The characteristics of a substance that are likely to ignite it in the presence of ignition sources. The properties of a material, such as flash point, flammable limits, vapor pressure, temperature relative to boiling point, autoignition temperature, and others.

ignition   The point at which the heating of something becomes self-perpetuating.

ignition temperature   The lowest temperature that will cause a gas/vapor to ignite and burn independent of the heating source.

IH   Industrial hygienist or industrial hygiene.

illness   A condition marked by pronounced deviation from the normal healthy state; the subjective state, or experience, of a person with disease.

illuminance   The amount of light falling on a surface. Illuminance is expressed in units of foot-candles or lux.

illumination   The amount or quantity of light, measured in foot-candles or lux, that is incident on a surface.

ILO   International Labor Organization.

IM   Intramuscular.

image (Microscope)   The optical reproduction of an object that is produced by a lens.

immediately dangerous to life or health (Respirator Use)   [IDLH]   An environmental condition that places a worker in danger. NIOSH considers an IDLH situation as the maximum contaminant concentration from which, in the event of respirator failure, one could escape within 30 minutes without a respirator and without experiencing any escape-impairing or irreversible health effects. The primary purpose for developing an IDLH was for respirator selection. The purpose for NIOSH establishing IDLH exposure concentrations was to ensure that a worker could escape from a given contaminated environment in the event of failure of the respiratory protective equipment being used.

immediately dangerous to life or health (OSHA)   An atmospheric concentration of any toxic, corrosive, or asphyxiant substance that poses an immediate threat to life or would cause irreversible or delayed adverse health effects or would interfere with an individual's ability to escape from a dangerous atmosphere.

immersion foot   *See* trench foot.

imminent danger (OSHA)   Any condition or work practice for which there is reasonable expectation that death or serious physical harm or permanent impairment of the health or functional capacity of employees could occur immediately or before the situation can be eliminated through normal enforcement procedures.

imminent health hazard   A significant danger to health that is considered to exist when sufficient evidence shows that a product, medication, practice, operation, or event results in a situation that requires immediate correction or cessation of an operation to prevent potential adverse health effects or potential injuries.

immiscible   Not capable of being uniformly mixed or blended.

immune   Not affected or responsive.

immune person   One who possesses specific protective antibodies or cellular immunity as a result of previous infection or immunization, or is so conditioned by such previous specific experience as to respond adequately with the production of antibodies sufficient to protect from illness following exposure to the etiologic agent of the disease.

immune response   A selective response of the vertebrate immune system whereby specific antibodies are produced in response to any of numerous substances, microorganisms, parasites, cells, etc. that are foreign to the body or that are interpreted as being foreign.

immune system   A body system that helps resist disease-causing germs, viruses, or other infections.

immunity   Insusceptibility. Biologically, immunity is usually to a specific infectious agent and is one result of infection. The quality or condition of being immune. An inherited, acquired, or induced condition to a specific pathogen. The power of the body to successfully resist infection and the effects of toxins.

immunity, acquired   Resistance to disease that has been acquired by an individual as a result of previous exposure to a natural pathogen or foreign substance.

immunity, active   Resistance developed in response to stimulus by an infectious agent or a vaccine and typically characterized by the presence of antibodies produced by the individual.

immunity, passive   Resistance to disease that has been acquired by the administration of an antibody-containing preparation (e.g., immune globulin).

immunization   The act of providing protection from a communicable disease by the administration of a living modified disease agent, a suspension of killed organisms, or an inactivated toxin.

immunoassay   The measurement of an antigen–antibody interaction.

immunodeficiency   A deficiency in immune response, either in that mediated by immune humoral antibody or in that mediated by immune lymphoid cells.

immunogenicity   The ability of an antigen to produce an immune response.

immunology   The medical study of immunity.

immunoprophylaxis   Prophylactic vaccination to increase antibody levels in the body to levels that are protective.

immunosuppressed   A condition in which the body's immune system does not function normally.

immunotoxin   An antibody to the toxin of a microorganism, zootoxin (spider or bee toxin), or phytotoxin (toxin from a plant) that combines specifically with the toxin, resulting in the neutralization of its toxicity.

impaction   The forcible contact of particles with a surface. The cascade impactor is a device that operates on this principle.

impact noise (OSHA)   Variations in the noise level such that the maximum noise level occurs at intervals of greater than one second.

impairment   The loss of function of an organ or part of the body compared to what previously existed.

impedance   A measure of the total opposition to current flow in an alternating current circuit.

impeller   The rotating blade of a fan that forces air in a given direction under pressure.

impermeable   Not capable of being permeated, or not allowing substances to pass through the openings or interstices of the material.

impervious (Protective Clothing)   Resistant to penetration by liquids or gases.

impetigo   A streptococcal infection of the skin characterized by vesicles that rupture, forming rapidly enlarging and spreading erosions.

impingement   The process by which particulate material in air is collected by passing the air through a nozzle or jet and impinging the air–particle mixture onto a surface that is immersed in a liquid such as water. The particles are retained in the liquid. The midget and Greenburg–Smith impingers are examples of instruments using this principle of dust collection.

impinger   A sampling device used to collect airborne particulates. The midget impinger and the Greenburg–Smith impinger are typical types.

implementation   The putting into effect of a definite plan and procedure for ensuring fulfillment and achievement of a given end.

implosion   A violent inward collapse of an item, such as an evacuated glass vessel.

impoundment   A body of water that is confined by a dam or other barrier.

impulse noise   An acoustic event characterized by very short rise time and duration.

impurity   The condition or quality of being polluted or contaminated.

in   Inch.

in²   Square inch(es).

in³   cubic inch(es).

incandescent   A material that is heated to the point that it becomes hot enough to radiate visible light.

incendiary device   A device that is capable of causing fire.

incendive spark   A spark of sufficient temperature and energy to ignite a flammable vapor/gas.

inch of mercury   [in Hg]   A unit used in measuring or expressing pressure. One inch of mercury pressure is equivalent to 0.491 pounds per square inch.

inch of water   [in H_2O]   A unit of pressure equal to the pressure exerted by a column of liquid water one inch high at standard temperature.

incidence   Number of new cases of disease within a specified period of time.

incidence rate (Epidemiology)   The ratio of the number of new cases of a disease during a time period to the total population at risk during that time period.

incidence rate (OSHA)   The number of injuries and/or illnesses or lost workdays per 100 full-time employees per year or per 200,000 hours of exposure.

incident (Safety)   An unplanned event that interrupts the completion of an activity and that may or may not include property damage or injury.

incineration (Air Quality)   The controlled process in which combustible solid, liquid, or gaseous wastes are burned.

incipient fire   *See* incipient stage fire.

incipient stage fire   A fire that is in the initial or beginning stage and that can be controlled or extinguished by portable fire extinguishers, Class II-type standpipe, or small hose systems without the need for protective clothing or breathing apparatus.

inclined manometer   A manometer, used in pressure measurement, that amplifies the vertical movement of the water column through the use of an inclined leg.

incombustible   Incapable of burning.

incompatible material   Material that, if mixed with another specified material, could result in an undesirable or dangerous reaction.

incoordination   Lack of normal adjustment of muscular motion.

incubation   Holding cultures of microorganisms under conditions favorable to their growth.

incubation period   In the development of an infection, it is the period of time from its entry or initiation within an organism to the first appearance of signs or symptoms.

indeterminate errors   Errors that occur randomly and whose cause is not determinable and thereby cannot be corrected for.

indicator tube   A tube containing a chemisorbent material that provides an indication, typically by color change, of the concentration of a specific substance in air drawn through the tube. Also referred as a detector tube.

indigenous   Occurring or living naturally in an area.

indirect ionizing particle   Particles that can liberate direct ionizing particles or can initiate a nuclear transformation. Examples include neutrons and gamma rays.

indolent   Person who is not inclined to work. An habitually lazy person.

indoor air   That part of the atmosphere or air that occupies the space within the interior of a house or building.

indoor air pollutant [IAP] Pollutants within a structure such as radon, carbon dioxide, asbestos, lead, pesticides, carbon monoxide, hydrocarbon vapors, and others, that may contribute to discomfort or disease.

indoor air quality [IAQ] The study, evaluation, and control of indoor air with respect to temperature, humidity, odor, airborne contaminants, cleanliness, and other indoor environmental factors that may affect the health or comfort of occupants.

induced draft Negative pressure created by the action of a fan or ejector located between a combustion chamber and a stack/exhaust vent.

induced radioactivity Radioactivity that is produced in a substance by a nuclear reaction.

induction Causing to occur.

induction period (Toxicology) The time between first exposure and the induction of a health effect at the molecular level.

industrial gases Gases that are customarily used in industrial processes, such as in welding and cutting, heat treating, refrigeration, water treatment, and others.

industrial hearing loss Hearing loss caused by work-related noise exposure.

industrial hygiene [IH] The science and art devoted to the anticipation, recognition, evaluation, and control of those environmental factors or stresses arising in or from the workplace that may cause sickness or impaired health and well-being or significant discomfort or inefficiency among workers or residents of the surrounding community.

industrial hygienist [IH] An individual who possesses a degree from an accredited university in industrial hygiene, engineering, chemistry, physics, medicine, or other physical or biological science, and who, by virtue of specialized studies and training, has acquired competence in industrial hygiene.

industrial radiography The examination of the macroscopic structure of materials by nondestructive methods using sources of ionizing radiation.

industrial safety A people-directed activity for preventing personal injuries as opposed to system problems.

industrial ventilation The methods and operation of equipment associated with the supply or removal of air, by natural or mechanical means, to control airborne contaminants resulting from industrial operations.

inebriation The condition of being drunk.

inert (Condition) The condition of a tank or other enclosure when the oxygen content of the atmosphere throughout the enclosed space has been reduced to 8 percent or less by volume through the addition of an inert gas.

inert dust Dust that has a long history of little or no adverse effect on lungs and does not produce significant organic disease or toxic effect when exposures are kept under reasonable control. Such dusts are called inert dusts (biologically). *See* nuisance dust.

inert gas A nonreactive gas such as argon, helium, neon, or krypton. These gases will not burn or support combustion and are not toxic. Nitrogen is often used as an inert gas in process operations for reducing the risk of fire/explosion.

inertia The tendency of a body to remain at rest or of a body in motion to stay in a straight line unless disturbed by an external force.

inertial separator (Ventilation)   An air cleaning method for the removal of particulates. Its effectiveness depends upon the inability of particulates to make sharp turns due to their inertia being greater than the air/gas stream transporting them. Thus, they (particularly the larger ones) are removed from the air stream by centrifugal force.

inerting   The act of displacing a flammable/combustible atmosphere in a space by a noncombustible gas to such an extent that the resulting atmosphere is not a fire or explosion hazard.

inert ingredient   An inactive ingredient that serves as a filler or binder.

infarct   A necrotic area of tissue that is the result of failure of blood supply to the area.

infected person   Sick person or patient with apparent infection and who is a carrier. A person or animal that harbors an infectious agent and that has either manifest disease or inapparent infection.

infection   Invasion by pathogenic organisms, as well as their development or multiplication in a body part, in which conditions are favorable for growth, production of toxins, and subsequent injury to tissue.

infectious   Capable of invading a susceptible host, replicating, and causing an altered host reaction, such as disease.

infectious agent   An organism that is capable of producing infection or infectious disease.

infectious disease   A clinically manifest disease of man or animal resulting from an infection.

infectious material   Human body fluid, including amniotic fluid, that is contaminated with blood, cerebrospinal fluid, semen, vaginal secretions, synovial fluid, pleural fluid, pericardial fluid, peritoneal fluid; any body fluid that is difficult or impossible to differentiate; any unfixed tissue or organ from a human; HIV- or HBV-containing cultures or culture medium; and blood, organs, or other tissues from experimental animals infected with HIV or HBV.

infectious waste   A waste material that is suspected by a health care professional in charge of being capable of producing an infectious disease because (1) it has been or is likely to have been contaminated by an organism likely to be pathogenic to healthy humans; (2) the organism is not routinely and freely available in the community, or (3) the organism has a significant probability of being present in significant quantities and with sufficient virulence to transmit disease. Any item that has been in contact with infectious organisms or with materials such as blood, serum, excreta, tissue, etc. that may be infected must be considered infectious unless it has been treated.

infectivity   The state or quality of being infectious.

inference   The process of passing from observations to generalizations. A conclusion based on a premise.

inferential statistics   A technique for inferring something and drawing conclusions from data or information obtained from a representative sample taken from a population. Provide a means of drawing conclusions about a larger body or population based on sample data from that population.

infestation   The lodgment, development, and reproduction of arthropods or other pests on the surface of the body or in the clothing.

infiltration   Air leakage into a building through cracks, ceilings, walls, floors, and small openings due to the existence of positive pressures outside the building surface.

infirmity   A weak state of the body or mind as a result of any condition that produces such weakness.

inflammation   The reaction of the body to injury, whether by trauma, infection, or chemical irritation.

infrared detector   A measurement technique in which infrared radiation (IR) is passed through a cell containing the sampled material. The absorption of the infrared energy at a wavelength which coincides with the absorption band of the analyte (contaminant) is proportional to the amount of contaminant present. This principle can also be applied to the determination of materials present in air drawn through a cell through which a beam of IR radiation is passed.

infrared radiation   [IR]   Electromagnetic energy with wavelengths from 760 nanometers to 14,000 nanometers. This wavelength is longer than the visible spectrum, but shorter than the radiofrequency range. The IR spectrum is divided into the IR-A, IR-B, and IR-C regions. The IR-A region is of most concern from an industrial hygiene standpoint.

infrasonic   At a frequency below the audio frequency range. Also called subsonic.

infrasound   Low frequency sound that cannot be heard.

ingestion   The process of taking substances into the body by mouth.

inhalable fraction   The mass fraction of total airborne particulates that is inhaled through the nose and mouth.

inhalation   The breathing in of a substance, such as air or a contaminant in the atmosphere.

in Hg   Inch(es) of mercury.

inhibitor   A substance used to retard or halt an undesirable reaction, such as the addition of a chemical in small quantities to an unstable material to prevent a vigorous reaction.

in H_2O   Inch(es) of water.

injection well   A well into which a fluid (i.e., liquid or gas) is injected.

injury   Physical damage to body tissues caused by an accident or by exposure to environmental stressors.

injury, disabling   An injury in which the injured is unable to come to work for at least one full shift following the injury.

innage   The height of a liquid in a tank from the bottom datum plate of the tank to the liquid surface.

inner ear   That part of the ear that converts vibration to nerve impulses. The location of the sensory receptors responsible for the initiation of neural impulses in the auditory nerve.

innervate   To supply nervous energy or nerve stimulus to a part of the body.

innocuous   Harmless, or having no adverse effect.

inoculation   The introduction of the virus of a disease or other antigenic material into the body to immunize, cure, or experiment. Also the implantation of microorganisms or infectious material into a culture medium.

inoculum   The material used in an inoculation.

inorganic compound   Compound that does not contain carbon as the principal element, as distinguished from an organic compound that does.

insect   Any of numerous usually small invertebrate animals having an adult stage characterized by three pairs of legs, a segmented body, and usually two pairs of wings. Examples are the spiders, centipedes, flies, ticks, and others.

insecticide   A substance or mixture of substances for preventing or inhibiting the reproduction, development, and growth of insects, or destroying or repelling them.

insertion loss (Acoustics)   The difference in sound pressure level at a fixed measuring location before and after a noise control has been applied to a noise source.

insidious   Spreading in a subtle manner.

in situ   In its original place.

insolation   The solar radiation incident on the earth.

insoluble   Incapable of being dissolved.

insomnia   Chronic inability to sleep.

Inspirable Particulate Mass   [IPM]   Particulates that are hazardous when deposited anywhere in the respiratory tract.

Inspirable Particulate Mass TLVs   [IPM-TLVs]   Exposure limits that are applied to those materials that are hazardous when deposited anywhere in the respiratory tract.

inspirate   Inhaled gas or air.

Instrument Society of America   [ISA]   A group that sets standards of performance for instruments made and used in the U.S.

integrated circuit   A small chip of silicon on which miniaturized circuits have been etched.

integrated sample   The sum of a series of small samples or continuous samples over a period of time used to create a large average sample.

intensity   The energy transferred by a wave per unit time across a unit area perpendicular to the direction of propogation.

intensity (Illumination)   The quantity of light a source gives off in a given direction.

interference (Sampling)   An undesired positive or negative response caused by a substance other than the one being monitored. Substances that may be present in the atmosphere along with the contaminant of interest, which, when sampled, affect the reading of an instrument, detector tube, or the analysis of the sample. Interferences can be positive or negative, significant or insignificant, accounted for or unaccounted for, and generally must be considered when assessing an exposure situation.

interference equivalent   Mass or concentration of an interfering substance that gives the same measurement reading as a unit mass or concentration of the substance being measured.

interlock   An electrical or mechanical device for preventing the continued operation of an instrument if the interlock is not working, or for inactivating an instrument/appliance until a condition has been corrected to enable its safe operation.

intermediate   A chemical formed as a middle step in a series of chemical reactions, especially in the formation of organic compounds.

intermittent noise   Noise which occurs intermittently or falls below the audible or measurable level one or more times over a given period. Also defined as an exposure to a given broadband sound-pressure level several times during a normal work day.

internal contamination (Ionizing Radiation)   Radioactive contamination within a person's body as a result of inhaling, swallowing, or skin puncture by radioactive materials.

internal conversion   A mechanism of radioactive decay in which transition energy is transferred to an orbital electron, causing its ejection from the atom.

internal duct liner (Indoor Air Quality)   Insulating material that is generally porous, which is positioned inside ductwork as noise and/or thermal insulation. It can become a strong source of indoor microbiological pollutants if it becomes dirty and moist.

internal radiation   An ionizing radiation dose from a radioactive source as a result of its introduction into the body by inhalation, ingestion, puncture wound, or other means.

internal respiration   The exchange of gas between tissue blood and the tissue cells.

International Agency for Research on Cancer   [IARC]   A agency of the World Health Organization that is one of three sources that OSHA refers to for data on a material's carcinogenicity.

International Committee on Radiation Protection   [ICRP]   An international group of scientists who develop recommendations on ionizing radiation dose limits and other radiation protection measures.

International Society of Indoor Air Quality and Climate   [ISIAQ]   A group that addresses the science and technology of indoor air quality issues through meetings and publications.

International Union on Pure and Applied Chemistry   [IUPAC]   A voluntary, nonprofit organization representing chemists in about 50 countries with the object of facilitating international agreement and uniform practices in both the academic and industrial aspects of chemistry. In addition, this organization carries out research on food technology, water and air quality, and in other fields.

interpolate   To determine the value of a function that is between two known values.

interrogatories (Legal)   A series of questions in writing that are submitted to one party for developing a written response to each that may then be used in a legal proceeding.

interstitial   The space between cellular components or parts of a structure or organ.

intertrigo   Superficial dermatitis caused by moisture, warmth, friction, and sweat retention. Obesity is a predisposing factor.

intoxication   The inducing of any of a series of progressively deteriorating states ranging from exhilaration to stupefaction.

intracellular   Within a cell or cells.

intractable   Resistant to change.

intramuscular   Within a muscle.

intraperitoneal   [IP]   Within the abdominal-pelvic cavity.

intrapleural   Within the chest cavity.

intratracheal   Endotracheal, or within or throughout the trachea.

intravenous   Within a vein or veins.

intravenous drug   A drug that is injected by needle directly into a vein.

intrinsic   Situated in or denoting internal origin and belonging solely to a body part.

intrinsically safe   Equipment that is incapable of releasing sufficient thermal or electrical energy, under normal or abnormal conditions, to cause ignition of a specific hazardous atmospheric mixture. Abnormal conditions may include accidental damage to any part of the equipment or wiring, insulation, or other failure of electrical components, maintenance operations, and other similar conditions.

invasive   The ability of a microorganism to enter the body and spread throughout the tissues, as well as the ability to enter the body and destroy surrounding tissue.

inverse square law   A law that states that doubling the distance from a source of light, noise, ionizing radiation, etc. reduces the intensity of the exposure to the source at that point by one fourth (i.e., the intensity varies inversely with the square of the distance from the source).

inversion   A meteorological condition that occurs when higher air is warm and, as a result, prevents cooler air near the surface of the earth from rising. As a consequence, contaminants cannot disperse into the atmosphere effectively.

invertebrate   An animal that does not have a backbone.

inverter   A device for converting direct current to alternating current.

inviolable   Safe or secured against violation and impregnable to trespass.

in vitro   Within glass. Observable in a test tube or in an artificial environment.

in vivo   Within the living body. An experiment or study that is conducted in a living organism.

involuntary muscle   A muscle that is not under the control of the will.

involuntary risk   A risk that an individual has not consented to accept.

in wg   Inch(es) water gauge.

IOHA   International Occupational Hygiene Association

ion   An atom that has lost or gained one or more electrons and is left with a positive or negative electrical charge.

ion exchange   The reversible interchange of ions of like charge between an insoluble solid and a surrounding liquid phase in which there is no permanent change in the structure of the solid.

ion exchange resin   Synthetic resin that contains active groups, enabling the resin to combine with or exchange ions between it and those in another substance.

ionization   The process by which a neutral atom or molecule acquires a positive or negative charge by adding electrons to or removing electrons from atoms or molecules.

ionization chamber (Ionizing Radiation)   Device designed to measure the quantity of ionizing radiation in terms of the charge of electricity associated with ions produced within a defined volume.

ionizing radiation   The emission of electromagnetic or particulate radiation that is capable of producing ions, directly or indirectly, by interaction with matter. Any type of radiation that is capable of producing ionization in materials it contacts. In biological systems, such radiation must have a photon energy greater than 10 eV.

ionosphere   A layer of the earth's atmosphere that is electrically conductive because of the ionization of rarefied atmospheric gases by incident solar radiation.

ion pair   A positively charged ion and an electron. One method by which ionizing radiation gives up its energy is by the production of ion pairs.

IP   Intraperitoneal.

IPM TLVs   Inhalable Particulate Mass TLVs.

IR   Infrared radiation.

iritis   Inflammation of the iris.

IRPA   International Radiation Protection Association.

irradiance   Radiant flux per unit area that is incident upon a given surface.

irradiation   Exposure to ionizing radiation or any other kind of radiation, such as heat, light, etc.

irrespirable   Unfit for breathing.

irreversible effect   An effect that is not reversible once the exposure has terminated.

irreversible injury   An injury that is neither repairable nor expected to heal.

irritant   An external stimulus that produces a response in a living organism. A substance that produces an irritating effect when in contact with skin, eyes, or the respiratory system. A primary irritant is one that has been found to produce an irritating effect at the location where contact occurs.

irritant smoke   A smoke-like material that is used in determining whether a mechanical air-purifying respirator wearer achieves a good fit in a qualitative fit test. Stannic oxychloride or titanium tetrachloride is used as the source of irritant smoke. This smoke is also used in ventilation system evaluations, (i.e., smoke tubes).

irritating material   A liquid or solid substance that causes reversible inflammatory effect on human tissue at the site of contact, or a liquid that has a severe corrosion rate on steel or aluminum.

irritation   An inflammatory response or reaction of tissues resulting from contact with a material.

ISA   Instrument Society of America.

ischemia   A localized reduction of blood supply to a part of the body.

ISIAQ   International Society of Indoor Air Quality and Climate.

ISO   International Standards Organization.

isoamyl acetate   A liquid chemical having a banana-like odor that is used as a test agent in qualitative fit testing of respirators.

isobar   Line on a weather map or chart connecting points of equal atmospheric pressure.

isocyanate asthma   Bronchial asthma that is a result of an allergy to toluene diisocyanate and similar cyanate compounds.

ISO 9000   A worldwide quality management system standards developed by the ISO to provide a systematic approach to process management.

ISO 14,000   A voluntary international standard of tools and systems for environmental management which were developed by the ISO.

isokinetic sampling   Sampling for particulates at a rate that establishes a velocity at the sampling probe inlet equal to that of the airstream from which the sample is being collected.

isolation (Acoustics)   The use of materials or construction around a noise source to limit the transmission of sound from that source.

isolation (Communicable Diseases)   The separation for the period of communicability of infected persons or animals, in such places and under such conditions as to prevent or limit the direct or indirect conveyance of the infectious agent from those infected persons to other persons who are susceptible or who may spread the agent to others.

isolation (Confined Space)   A procedure whereby a confined space is removed from service and completely protected against inadvertent release of material by blanking off with a skillet-type metal blank between flanges, a double block and bleed system, electrical lockout of all sources of power, and blocking or disconnecting all mechanical linkages.

isolation (Process)   A procedure by which a space is completely protected against a release of energy or material into the space by such means as blinding, removing a section of pipe, installing a double block and bleed blinding system, etc.

isomer   One of a group of compounds having the same number and kind of constituent atoms, thus the same molecular weight, but different molecular structures.

isometric   Exhibiting equality in dimensions.

isopleth   *See* sound level contours.

isotherm   Line on a weather map or chart connecting locations that have an identical mean temperature for a given period or a given time.

isothermal   Of, pertaining to, or indicating equal temperatures.

isotopes   Nuclides having the same number of protons in their nuclei and therefore the same atomic number. Isotopes of the same element have essentially identical chemical properties, but different nuclear properties.

Itai Itai disease   A disease that was considered to be a result of eating rice that had been contaminated with cadmium from industrial emissions.

IUPAC   International Union of Pure and Applied Chemistry.

I.V.   Intravenous.

J

J   joule(s), 1 E^7 ergs.

jackhammer   A handheld pneumatic machine for drilling through and breaking concrete/rock.

jamb   *See* airfoil.

jargon   Communication that includes unfamiliar words, abbreviations, and permutations of normal terms, usually within the context of a special occupation, such as industrial hygiene and safety.

jaundice   A syndrome characterized by the deposition of bile pigment in the skin and mucous membranes, with resulting yellow appearance of the person.

Jaws of Life   A trademark for a pneumatic tool consisting of a pincerlike metal device that is inserted into the body of a severely damaged vehicle and energized to provide access to a person trapped inside.

J.D.   Doctor of Laws.

jeopardize   To expose to loss or injury, or make vulnerable or precarious.

jet stream   A high-speed wind near the troposphere, generally moving in an easterly direction, often at speeds up to 250 miles per hour.

JHA   Job hazard analysis.

job hazard analysis   [JHA]   A procedure for identifying all the hazards or potential accidents/hazards associated with each step or task of a job assignment and developing solutions that will eliminate the hazards and prevent accidents.

job safety analysis   [JSA]   A procedure used to review job methods to discover hazards due to work-station design or changes in job requirements. After hazards have been identified, control measures are developed.

joint (Anatomy)   A point at which two separate bones are joined together by ligaments.

joint (Smoking)   Any cigarette made from marijuana.

joule   [J]   International System unit of energy, equal to the work done when a current of 1 ampere is passed through a resistance of 1 ohm for 1 second.

JSA   Job safety analysis.

judicious   Having or exhibiting good judgment.

jugular vein   Any of the large veins in the neck.

junction box   An enclosed panel used to connect or branch electric circuits without making permanent splices.

jury-rigged   Rigged for temporary use.

juxtapose   Placed together or situated side by side.

K

K   Kelvin.

kaolin   An aluminum silicate clay used in refractories and ceramics, as well as a filler or coating for paper and textiles.

kaolinosis   A pneumoconiosis resulting from the inhalation of kaolin clay dust.

katharometer   Instrument used for the electrometric determination of basal metabolic rate.

kB   Kilobyte(s), 1 E³ B.

kcal   Kilocalorie(s), 1 E³ cal.

kCi   Kilocurie(s), 1 E³ Ci.

keloid   A mass of fibrous connective tissue, usually at the location of a scar, or the formation of excessive collagen during connective tissue repair.

Kelvin   [K]   A temperature scale in which the zero point is −273°C. Also referred to as the absolute temperature scale.

keratin   Tough, fibrous protein containing sulfur and forming the outer layer of epidermal structures, such as hair, nails, etc.

keratitis   Inflammation of the cornea.

keratosis   Any horny growth on the skin, such as a wart.

kerogen   The organic component of oil shale that yields oil when the shale is heated.

keV   Kilo electron volt(s), 1 E³ eV.

kg   Kilogram(s), 1 E³ g.

kHz   Kilohertz, 1 E³ Hz.

kilo   [K]   One thousand.

kilocalorie   A unit of heat equal to one thousand calories.

kilo electron volt   [keV]   One thousand electron volts.

kilogram   [kg]   One thousand grams.

kilohertz   [kHz]   One thousand hertz.

kilometer   [km]   A unit of distance equal to 1,000 meters.

kilopascal   A unit of pressure equal to one thousand pascals. One pound per square inch (psi) of pressure is equivalent to 6.894757 kilopascals.

kilovolt   [kV]   One thousand volts.

kilowatt   [kW]   One thousand watts.

kilowatt–hour   Use of 1 kilowatt of electrical energy for 1 hour.

kinematics   The study of motion, exclusive of the influence of mass and force.

kinesiology (Occupational)   The study of the motion of the human body and its limitations.

kinetic energy   Energy possessed by a mass because of its motion.

km   Kilometer(s), 1 E³ m.

knot   A unit of speed equal to 1 nautical mile per hour.

Kohler illumination   A type of illumination used in microscopy in which the light source is imaged in the aperture of the system, and the lamp condenser is imaged in the specimen plane, in order to obtain even brightness in the field of view and optimum resolving power of the microscope system.

kPa   Kilopascal(s), 1 E³ Pa.

kPa abs   Absolute pressure in kilopascals.

kV   Kilovolt(s), 1 E³ V.

kVp   Kilovolt peak.

kW   Kilowatt(s), 1 E³ W.

kwashiorkor   Severe malnutrition, characterized by anemia, edema, skin depigmentation, and loss of hair or change in hair color.

kWh   Kilowatt–hour.

kyphosis   Curvature of the spine of the upper back. Also referred to as hunchback.

L

L   Liter(s).

label   The printed or graphic material on or attached to a container, such as a box, bottle, can, or other type container or its wrapper.

labile   Chemically unstable.

lacrimation   The excessive secretion and discharge of tears.

lacrimator   A substance, such as a gas, that increases the flow of tears.

LADA   Laboratory animal dander allergies. Individuals who are allergic to animal dander are classified as LADA.

lading   The contents, such as is shown on a bill of lading.

lagging (Acoustics)   An acoustical treatment involving the encapsulation of vibrating structures or ducts containing fluid-borne noise in order to reduce radiated noise.

lag time (Instrument)   The time required for the first observable change in instrument response due to a sudden change in input concentration. *See* dead time.

lambert   [L]   A unit of luminance in the CGS system.

laminar air flow   Airflow in which the air within a designated space moves with uniform velocity in one direction and along parallel lines. Also referred to as streamline flow.

laminar flow clean room   A room with laminar air flow and Class 10,000 Clean Room or better.

lamp   Man-made light source.

land breeze   Air movement onshore during the daytime when the land mass becomes warm relative to the air over the sea. As a result, onshore air circulation develops.

land-farming   The application of waste to land and/or incorporation of the waste into the surface soil, including the use as a fertilizer or soil conditioner.

landfill   A disposal facility where waste is buried in layers below the ground, then compacted and covered.

LANL   Los Alamos National Laboratory (previously called Los Alamos Scientific Laboratory, LASL).

lanyard   A short rope suitable for supporting one person or for securing rigging.

lapse rate   The rate of decrease in temperature with increasing height or altitude.

larvicide   An agent that destroys or incapacitates larvae.

laryngeal   Of or pertaining to the larynx.

laryngitis   Inflammation of the larynx.

larynx   The essential sphincter guarding the entrance into the trachea and functioning secondarily as the organ of voice.

laser   Acronym for light amplification by stimulated emission of radiation. Lasers are devices that convert electromagnetic radiation of mixed frequencies to one or more discrete frequencies of highly amplified and coherent visible radiation. *See* Class for the classification of lasers.

laser beam   An intense beam of light that spreads out very little over its path of use.

laser safety officer   One who has authority to monitor and enforce measures to control laser hazards and to carry out evaluations to identify measures to control laser hazards.

laser system   An assembly of electrical, mechanical, and optical components, including a laser.

lasing medium (Lasers)   The material that absorbs and emits the laser radiation. Lasers can be classified according to the state of their lasing media (e.g., gas, liquid, solid, and semiconductor).

LASL   Los Alamos Scientific Laboratory (renamed Los Alamos National Laboratory, LANL).

lassitude   Weakness or exhaustion.

latency period (Carcinogenesis)   The period of time between the onset of exposure to a carcinogen and the clinical detection of resulting cancers.

latency period (General Diseases)   A seemingly inactive period between exposure to an injurious agent and the first manifestation of response to it. *See* also latent period.

latent   Present or potential, but not manifest.

latent heat of fusion   The heat required to convert a unit mass of solid to a liquid at the melting point.

latent heat of vaporization   The heat required to convert a unit mass of substance from the liquid state to the gaseous state at a given pressure and temperature.

latent period   The period of time between exposure to an injurious agent (chemical, physical, or biological) and the observation of an effect. The incubation period of an infectious disease. Also referred to as the latency period of a disease.

latex sensitivity   A response to the wearing of latex gloves and other items made of latex, giving rise to irritant contact dermatitis, allergic contact dermatitis, and direct sensitivity to latex.

lavage   The irrigation or washing out of an organ, such as the stomach or bowel.

lavatory   A room equipped with washing facilities and, usually, toilet facilities.

lb   Pound(s).

lb/ft³   Pounds per cubic foot.

LBP   Lead-based paint.

LBPPPA   Lead-Based Paint Poisoning Prevention Act.

LC   Liquid chromatography.

LC   Lethal concentration.

LC$_{50}$ Median lethal concentration.

LD   Legionnaire's disease.

LD   Lethal dose.

LD$_{50}$ Median lethal dose.

L$_{dn}$ Average day–night sound level.

LDAR program   Leak detection and repair program for fugitive emission sources.

leach   A process in which materials in the soil move into a lower layer or are dissolved and carried through the soil by water.

leachate   A liquid that has percolated through soil, wastes, or other media, and has removed materials.

lead-based paint (General)   [LBP]   A paint or other surface coating product that has a lead content of 0.06 percent by weight in the total nonvolatile content of the paint, or by weight in the dried paint film.

lead-based paint (EPA)   Paint or other surface coatings that contain lead equal to or in excess of 1.0 milligram per square centimeter or 0.5 percent by weight. Some jurisdictions consider lead-based paints to be those having 0.06 percent or more of lead by weight.

lead-containing paint   See lead-based paint.

lead encephalopathy   An irreversible, degenerative disorder of the brain caused by the presence of lead. It produces coma, convulsions, and delirium, also behavioral changes such as insomnia, restlessness, loss of memory, irritability, confusion, and hallucinations.

lead line   A symptom of lead poisoning. A blue line on the gums as a result of excessive exposure to lead.

lead paint hazard   The presence of lead-based paint in places that would pose a danger to people if ingested or inhaled.

lead poisoning   The result of absorbing too much lead into the body. Symptoms include weight loss, stomach cramps, irritation, anemia, convulsions, and mental depression, and affects the brain, nervous system, blood, and digestive system.

leakage (Ionizing Radiation–Sealed Source)   That which occurs when some of the radioactive material contained in a sealed source penetrates the housing while in storage or use, and is therefore out of the sealed source, posing a greater hazard than if it were effectively contained.

leak test (Ionizing Radiation)   A type of test for determining whether a radioactive material is effectively contained or has escaped, or leaked, from a sealed source. It involves wiping surfaces on which the material would collect if it was released from the sealed source.

least fatal dose   The smallest dose of a substance that can cause death to a specified type of organism.

leeward   Downwind.

Legionnaire's disease   [LD]   Pneumonia caused by a bacterium, *Legionella pneumophila*. It has occurred among occupants of buildings in which this organism is present in the air at high concentrations.

Legionella   The bacterium that is the causative agent of Legionnaires' disease and Pontiac fever.

legionellosis   Diseases caused by Legionella bacteria.

LEL   Lower explosive limit.

leptospirosis   An infection transmitted to man by dogs, swine, and rodents, or by contact with contaminated water.

lesion   Any pathological or traumatic discontinuity of tissue or loss of a part.

LET   Linear energy transfer.

lethal   Sufficient to cause, or capable of causing, death.

lethal concentration 50 percent   [LC 50]   The concentration of a substance that will kill 50 percent of the animals exposed to it.

lethal dose   The dose of a substance necessary to cause death.

lethal dose 50 percent   [LD 50]   The single dose of a material which, on the basis of laboratory tests, is expected to kill 50 percent of a group of test animals.

lethality   The property, quality, or state of being lethal.

lethargy   Sluggish indifference, or a state of unconsciousness resembling deep sleep.

leukemia   A progressive malignant disease in which there is an overproduction of white blood cells, or a relative overproduction of immature white blood cells and enlargement of the spleen.

leukemogenic   A substance that can cause leukemia. Also referred to as a leukomogen.

leukocyte   White blood cell or corpuscle.

leukocytosis   A transient increase in the number of white cells in the blood (leukocytes) as a result of fever, infection, inflammation, etc.

leukoderma   Permanent loss of skin pigmentation.

leukopenia   A reduction to below the normal level of the number of white cells in the blood.

leukotoxic   Poisonous to leukocytes.

LFL   Lower flammable limit.

Lfm   Linear feet per minute.

LIA   Laser Institute of America.

liability   The state of being bound or obliged by law to do, pay, or make good something.

lice   General name for various parasitic insects that can infest man and cause diseases such as typhus, relapsing fever, and possibly the plague.

license   An authorization granted by a government agency to conduct an activity under the conditions specified in the license.

licensed material (Ionizing Radiation)   Source material, special nuclear material, or byproduct material that is received, possessed, used, or transferred under a special license issued by the licensing agency (e.g., NRC, Energy Research and Development Administration, or an agreement state).

licensee (Ionizing Radiation)   The company or person authorized to use a radioactive material obtained under a license issued by the NRC or an Agreement State.

life expectancy   The average length of time an individual may live based on analysis of actual time until the death of a large group of similar people.

lifeline   A rope, suitable for supporting one person, to which a lanyard or safety belt is attached.

life safety code   Guidance developed by the NFPA that is concerned primarily with life safety in case of fire in buildings for various occupancies. It addresses means of egress, signage, lighting, fire control of egress routes, construction, as well as alarms and fire drills.

life safety evaluation   A written review dealing with the adequacy of life safety features of a facility with respect to fire, storm, collapse, crowd behavior, and other safety-related considerations.

lifetime risk   A risk that could manifest anytime during the lifetime.

ligament   A band of fibrous tissue that connects bones or cartilages, serving to support and strengthen the joint.

ligand   A molecule, ion, or atom that is attached to the central atom of a coordination compound, a chelater, or other complex. Ligands are also referred to as complexing agents.

light field microscopy   Microscopic technique that relies on the amplitude modulation of light to make specimens visible. Different portions of the specimen absorb light to a differing degree, thereby providing specimen details as differences in the intensity of light reaching the eye.

light meter   A photometer with a light-sensitive layer that is used to measure illumination levels.

lightning arrester   A protective device, such as a rod or mast, for intercepting a lightning discharge before it can strike the object protected, and then discharging the lightning current harmlessly to the ground.

limit of detection [LOD]   The smallest amount of an analyte that can be distinguished from background or the lowest concentration that can be determined to be statistically different from a blank. Typically, it is that amount of analyte which is three standard deviations above the background response. *See* also lower detectable limit and detection limit.

limit of quantitation [LOQ]   The amount of analyte above which quantitative results may be reported with a specific degree of confidence. Typically, this value is 10 times the standard deviation of concentrations very near the limit of detection.

limnology   The scientific study of the life and phenomena of lakes, ponds, and streams.

linear   Describing a straight line or lines.

linear energy transfer (Ionizing Radiation) [LET]   The average energy that is locally imparted to a substance by a charged particle of given energy.

linear feet per minute [Lfm]   A measurement unit for air velocity.

linear hypothesis (Ionizing Radiation)   An assumption that a dose–effect curve in the high dose and high dose-rate ranges may be extrapolated through the low dose and low dose-rate ranges to zero, implying that, theoretically, any amount of radiation will cause some damage.

linearity (Instrument)   The quality describing the response characteristic of the instrument remaining constant with input level. The maximum deviation, as a percentage, between the instrument reading and the reading predicted by the line of best fit for the data.

linear range (Instrument)   The ratio of the largest concentration to the smallest concentration within which the detector response is linear. It is also expressed as the range (i.e., lower value to upper value) over which the detector response is linear.

linear response   A response that is directly proportional, within limits, to the dose received.

lipid pneumonia   A chronic condition resulting from the aspiration of oily substances into the lungs.

liquefaction   The changing of a solid into a liquid.

liquefied compressed gas   A compressed gas that is partially liquid at the cylinder pressure and a temperature of 70.

liquefied natural gas   A mixture of materials composed of carbon and hydrogen, with the principal components being ethane, propane, and butane.

liquefied petroleum gas   A mixture of material all composed of carbon and hydrogen with the principal components being butane with smaller amounts of ethane, ethylene, isobutane, and butylenes.

liquid air   Air that has been compressed and cooled to a point at which it liquefies.

liquid spiking   Introducing a solvent containing the analyte of interest directly onto a sorbent media. Subsequent desorption and analysis of the liquid spike should have a recovery of greater than or equal to 75 percent.

liter   A metric unit of volume that is one-thousandth of a cubic meter and 1.057 quarts.

lithosphere   The solid part of the earth, as distinguished from the hydrosphere and atmosphere.

live room   A room characterized by a small amount of sound absorption. *See* reverberation.

LLNL   Lawrence Livermore National Laboratory.

lm   Lumen.

ln   Natural logarithm.

LNG   Liquefied natural gas.

loading rack   Structure that permits loading of tank trucks and tank cars at top openings on the transport vehicle.

LOAEL   Lowest-observed-adverse-effect level.

local effect   An effect which occurs to a localized part of the body, such as irritation of the respiratory tract, eyes, or other area of the body with which the toxicant has come into contact.

local exhaust ventilation   A ventilation system, usually comprised of a hood or hoods, ductwork, possibly an air cleaner, and a fan. The system is designed to capture or contain contaminants at the source of generation for removal from the work environment. The mechanical removal of contaminants from a point of operation prevents their release into the work area.

local exhaust system   A system composed of an exhaust opening, such as a hood, ductwork to transport exhausted air to a source of suction (fan, eductor, etc.), and frequently, an air cleaner to remove contaminants from the exhaust air before discharge to the environment. The air cleaner is typically positioned before the fan in the system in order to prevent fan wear.

localized   Restricted to a limited region or to one or more spots.

local toxic effect   An effect that is observed at the site of contact. For example, a skin burn that results from contact with a corrosive substance.

lockjaw   *See* tetanus.

lockout device   A device that uses a lock and key to hold an energy-isolating device in the safe position for the purpose of protecting personnel. The device typically has the capability of accepting several locks so that each craft person working on the equipment serviced by the isolating device that is being locked out can place his or her lock on it so that it cannot be energized until all locks have been removed.

lockout/tagout   A formal procedure for isolating equipment, machinery, or a process to prevent its unintentional operation during maintenance, servicing, or for any other reason. The equipment, etc., is first put into an energy-isolated state, and each individual who will work on the device/equipment/machine/etc. places his/her lock and/or tag on the electrical switch or other start-up means in order to keep the device/machine/etc. in a zero-energy state until the work is completed by each individual who has affixed his/her lock and/or tag to it. The policy and procedure related to this practice is to clearly and specifically outline the purpose, responsibility, scope, authorization, rules, definitions, and measures to enforce compliance.

LOD   Limit of detection.

LOEL   Lowest observed effect level.

log   Logarithm.

$\log_{10}$ Logarithm to the base 10.

log-normal distribution   The distribution of the logarithms of a random variable that has the property that the logarithms are normally distributed.

log-normally distributed variable   A variable in which the logarithm of the variable is normally distributed.

long-term delayed effects   Health effects that occur years after exposure to a toxic substance, such as asbestos, silica, ionizing radiation, and others.

long-term exposure   Continuous or repeated exposure of an individual to a substance or agent over a period of several years or a working lifetime.

long ton   A unit of material equal to 1.120 short tons or 2240 pounds.

loose contamination (Radioactivity)   Radioactive contamination that may become airborne if disturbed, and that may result in the potential for it entering the body by inhalation, or indirectly by ingestion, through the handling of objects on which the contamination exists.

loose-fitting facepiece   A respiratory inlet covering that is designed to form a partial seal with the face, does not cover the neck and shoulders, and may or may not offer head protection against penetration.

loose paint   Peeling, flaking, or chipped paint; paint over crumbling, cracking or falling plaster; or plaster with holes in it. Paint that is damaged in any manner such that a child can get paint from the damaged area.

LOQ   Limit of quantitation.

loss control   A systematic means of identifying, controlling, and ultimately eliminating accidents and their associated cost.

lost time injury   A work injury resulting in death or disability and in which the injured person is not able to work the next regularly scheduled shift.

lost time injury rate   The number of disabling injuries per million man hours worked.

lost workdays   The number of workdays (consecutive or not) beyond the day of injury or onset of illness that an employee was away from work or limited to restricted work activity because of an occupational injury or illness.

loudness   The intensive attribute of an auditory sensation, in terms of which sounds may be arranged on a scale extending from soft to loud. It depends primarily on the sound pressure of the stimulus, as well as on its frequency and wave form.

loudness level   The loudness level of a sound, in phons, that is numerically equal to the median sound pressure level, in decibels, relative to $2 E^{-4}$ microbar, of a free progressive wave of 1000 hertz presented to listeners facing the source, which, in a number of trials, is judged by the listeners to be equally loud.

louver   Panels used in hoods for distributing airflow at the hood face.

low altitude   Elevations less than 4000 feet above sea level.

lower detectable limit (Instrument)   [LDL]   The smallest concentration of the substance of interest that produces an output change in a reading of at least twice the noise level.

lower explosive limit   [LEL]   The lower limit of flammability or explosibility of a gas or vapor at ordinary ambient temperatures, expressed in percent of the gas or vapor in air by volume. Also called the lower flammable limit (LFL).

lower flammable limit   *See* lower explosive limit.

lowest-observed-adverse-effect level   [LOAEL]   The lowest dose of a chemical that, in an experiment, causes a biologically significant sign of toxicity.

lowest-observed-effect level   The lowest experimentally determined dose that produces a biologically significant effect.

LOX   Liquid oxygen.

LPG   Liquefied petroleum gas.

LPM   Liter(s) per minute.

LPN   Licensed practical nurse.

LSO   Laser safety officer.

lubricity   The ability of an oil or grease to lubricate.

lucid   Clear and easily understood.

LUFT   Leaking underground fuel tank.

lumbago   Nonspecific low back pain.

lumbar   Refers to the five vertebrae of the lower back between the thorax and the pelvis.

lumbar insufficiency   Fatigue, stiffness, or pain in the lower back.

lumen   [lm]   The luminous flux on 1 square foot of a sphere of 1 foot radius, with a light source of 1 candle at the center radiating uniformly in all directions.

luminaire   A complete light fixture, including the lamp and the parts to distribute the light, position the fixture, and connect the lamp to the power supply.

luminance   The amount of light emitted by or reflected from a surface. Luminance is expressed as candles per square meter or foot-lamberts.

luminary   An object that gives light.

luminescence   The emission of light, as in phosphorescence, fluorescence, and bioluminescence, by processes that derive energy from essential nonthermal sources, such as chemical, biochemical, or crystallographic changes, the motion of subatomic particles, or the excitation of atomic systems by radiation.

luminous   Emitting light.

LUST   Leaking underground storage tank.

lux   [lx]   Metric unit of illuminance equal to 1 lumen per square meter. One foot candle is equal to 10.76 lux. Dividing the lux value by 10 provides the approximate equivalent foot-candle value.

LV%   Liquid volume percent.

lx   Lux.

Lyme disease   A bacterial disease transmitted to man by a tick bite. Symptoms of Lyme disease, including a rash, headache, fever, tiredness, numbness and others, mimic those of many other diseases.

lymph   A clear transparent, watery, multipurpose liquid that contains white blood cells and some red blood cells.

lymphatic system   The interconnected system between tissues and organs by which lymph is circulated throughout the body.

lymphocytes   Mononuclear white blood cells.

lymphocytopenia   Reduction in the number of lymphocytes in the blood.

lymphoid   Resembling or pertaining to lymph or tissue of the lymphatic system.

lymphoma   Various abnormally proliferative diseases of the lymphoid tissue of the lymphatic system. A tumor of lymphoid tissue.

lysis   Destruction of cells by a specific antibody causing rupture of the cell membrane.

lysosome   A type of organelle that contains potent substances capable of digesting liquid food material.

m   Meter(s).

m   Milli, 1 E $^{-3}$.

m   Molal (Solution).

M   Mega, 1 E^6.

M   Molar (Solution).

m^2   Square meter(s).

m^3   Cubic meter(s).

M-85   Motor fuel with 15 percent gasoline and 85 percent methyl alcohol.

mA   Milliamp, 1 E $^{-3}$ A.

MAC   Maximum Allowable Concentration.

Mace   Aerosol mixture of lacrymators.

maceration   The softening of a solid by soaking in a liquid.

Mach number   The ratio of the speed of a moving element, such as a plane, to the speed of sound in the surrounding medium, such as air.

machine guard   A means for protecting the operator and other employees in the machining area from hazards, such as those created by point of operation, ingoing nip points, rotating parts, flying chips, sparks, etc.

macrobiota   Macroscopic organisms that occur in a particular ecosystem or specified area.

macrobiotic   Long-lived.

macro-meteorology   Meteorological characteristics of a regional area, such as part of a province, region, state, or a larger area.

macrophage   Any of the large, highly phagocytic cells occurring in the walls of blood vessels and in loose connective tissue.

macroscopic   Visible to the eye without the aid of a microscope.

MACT   Maximum achievable control technology.

macula   A discolored spot on the skin that is not elevated above the surface.

magazine   A building or storeroom where ammunition or explosives are stored.

magnetic resonance imaging   [MRI]   A diagnostic imaging technique employing magnetic and radiofrequency fields to produce an image of body tissue and monitor body chemistry noninvasively.

magnification   The number of times the apparent size of an object has been increased by the lens system of a microscope.

main (Ventilation)   A duct or pipe connecting two or more branches of an exhaust system to the exhauster or air-cleaning equipment.

make-up air   Clean air brought into a building/area from outside to replace air that has been exhausted by ventilation systems and combustion processes. Typically introduced through the general ventilation system.

malaise   A vague feeling of bodily discomfort.

malaria   An infectious febrile disease caused by protozoa transmitted to man by the bite of an infected mosquito.

malformation   An abnormal form or structure generally due to abnormal development.

malfunction   Failure in performance.

malignant   When applied to a tumor, the term denotes the characteristics of being cancerous and capable of undergoing metastasis. The term is used to describe tumors that grow in size and also spread throughout the body.

malingerer   An individual who feigns illness or another problem in order to get out of work or responsibility.

malnutrition   The lack of suitable food due to inadequate diet, or deficient digestion, absorption, or assimilation.

malpractice   Misconduct or lack of proper professional skill on the part of a professionally trained person, such as an engineer, physician, dentist, lawyer, or other professional, in doing his/her work.

mammalian   Pertaining to mammals.

mammals   Common name of vertebrates of the class Mammalia.

management of change   A policy ensuring that changes made in process technology or through modifications to equipment or raw materials, new equipment put into service, in-service equipment taken out of service, changes in catalyst or operating conditions, changes to operating procedures, etc. do not result in an unwanted release of a hazardous chemical that could adversely affect employees or others. Such changes need to be identified, reviewed, and approved prior to their implementation.

mandate   An authoritative command, order, or instruction to do or not to do something.

manhole   A hole in the top or side of a tank or other type vessel or tower through which a person can enter. Also referred to as a manway.

manifest (Hazardous Waste)   A form used to identify the quantity, activity, and composition, and the origin, routing, and destination of a hazardous waste during its transportation from the point of generation to the point of treatment, storage, or disposal.

man-made air pollution   Air pollution that results directly or indirectly from human activities.

man-made mineral fiber   [MMMF]   A fibrous material that is man-made as opposed to a naturally occurring fibrous material like asbestos. Man-made mineral fibers are used as substitutes for asbestos-containing materials. They include fibrous glass, mineral wool, refractory ceramic fibers, etc.

man-made ionizing radiation   Ionizing radiation produced by a man-made source, such as an X-ray machine.

man-made vitreous fiber   [MMVF]   Fibrous, amorphous, inorganic substances that are made primarily from rock, clay, slag, or sand. They include fibrous glass, mineral wool (rock and slag), and refractory ceramic fibers.

manometer   An instrument for measuring pressure. It is essentially a U-tube or an inclined type partially filled with a liquid, such as water, mercury, or a light oil, and constructed so that the displacement of the liquid is an indication of the pressure being exerted on it.

manufactured gas   A gaseous fuel made from various petroleum products or from soft coal.

manway   *See* manhole.

MAP   Model Accreditation Plan (Related to the accreditation of persons who inspect for the presence of asbestos, develop asbestos management programs, etc. under AHERA and ASHARA as they relate to public buildings).

margin of safety (Industrial Hygiene)   The ratio between an allowable exposure limit for a specific toxicant and the level at which adverse effects have been noted.

margin of safety (Regulatory)   The estimated ratio of the no-observed effect level to the level accepted in a regulation.

margin of safety (Safety)   The failure point or breaking strength of a piece of equipment, such as a rope, sling, cable, or fitting, divided by the sum of the maximum weight to be hoisted. Also referred to as the safety factor.

margin of safety (Toxicology)   A factor determined by dividing the no-observed effect level by the estimated daily human dose.

marker   A unique component of a mixture that can be monitored and used as an indicator of the presence of the mixture.

marsh gas   Gas(es), such as methane and hydrogen sulfide, that is generated from the decomposition of organic matter.

Martin's diameter   Length of the line that divides a particle into two equal areas.

maser   Acronym for microwave amplification by stimulated emission of radiation.

masking (Acoustics)   A process by which the threshold of audibility for one sound is raised by the presence of another (masking) sound.

masking (General)   The blocking out of a sight, sound, or smell by another.

mass   The quantity of matter contained in a particle or body, regardless of its location in the universe.

mass flow meter   An electrically heated tube and an arrangement of thermocouples used to measure the differential cooling caused by a gas (e.g., air) passing through the tube. The thermoelectric elements generate a voltage proportional to the rate of gas flow through the tube.

mass median aerodynamic diameter   [mmad]   The mass median diameter of spherical particles of unit density that have the same falling velocity in air as the particle in question.

mass median size   The size of a particle in a distribution of particles such that the mass of all particles larger than the median is equal to the mass of all smaller particles.

mass number   The number of protons and neutrons in the nucleus of an atom.

mass psychogenic illnesses   Term used in describing illnesses experienced by workers for which no definitive cause/source can be identified.

mass spectrography   An instrumental analytical method for identifying substances from their mass spectra.

matched control   *See* control, matched.

material safety data sheet   [MSDS]   A compilation of information on a particular substance, such as a chemical or mixture. The information presented includes the composition and physical properties of the material, fire and explosion hazard information, adverse health effects information, reactivity data, procedures for responding to a spill or leak, as well as special protection requirements and precautions.

mathematical model   A representation of a process or relationship in mathematical form in which equations are used to simulate the behavior of the process or relationship under study.

matter   Anything that has mass, occupies space, can be perceived by one or more senses, and makes up any physical body.

max.   Maximum.

maximally exposed individual   The individual with the highest exposure in a given population.

Maximum Allowable Concentration   [MAC]   The level of airborne contaminant or physical agent to which a worker could be exposed over an 8-hour workday without experiencing an adverse effect. This term has been replaced in the U.S. by Threshold Limit Value. It is still used in some other countries. It is referred to as the maximum permissible concentration in some countries.

maximum evaporative capacity (Heat Stress)   The maximum amount of sweat that can be evaporated from the body's surface under the environmental conditions that exist. The evaporation of sweat is limited by the moisture content of the air.

maximum permissible concentration (Ionizing Radiation)   [MPC]   Recommended maximum average concentration of radionuclides in air or water to which a person (radiation worker or member of the general public) may be exposed, assuming 40 hours per week exposure for the worker and 168 hours per week for the public. That amount of radioactivity per unit volume of air or water which, if inhaled or ingested over a period of time, would result in a body burden that is believed will not produce significant injury.

maximum permissible dose (Ionizing Radiation)   [MPD]   The dose of ionizing radiation that, in the light of present knowledge, is not expected to cause significant bodily injury to a person at any time during his/her lifetime.

maximum permissible lift   [MPL]   Three times the acceptable lift in kilograms or pounds.

maximum use concentration (Respiratory Protection)   [MUC]   The maximum concentration that can exist for which a specific type of respiratory protection can be used. It is equal to the permissible exposure limit for the substance to which exposure occurs times the assigned protection factor.

Maxwell   A unit of magnetic flux in the meter–kilogram–second electromagnetic system.

mb   Millibar, 1 E^{-3} bar.

Mb   Megabyte, 1 E^6 b.

MBO   Management by objective.

mBq   Millibecquerel, 1 E^{-3} Bq.

MCA   Manufacturing Chemists Association.

MCEF   Mixed cellulose ester filter.

Mcf   Thousand cubic feet.

mcg   Microgram.

mCi   Millicurie(s), 1 E^{-3} Ci.

MCS   Multiple chemical sensitivity.

MDL   Method detection limit.

mean   The arithmetic average of a set of data.

mean free path (Acoustics)   The average distance sound travels between successive reflections in an enclosure.

mean radiant temperature   [MRT]   The temperature of a black body that would exchange the same amount of radiant heat as a worker would at the same location in a hot environment.

means of egress   A continuous and unobstructed way of exit from any point in a building to a public way. It consists of three separate parts: the exit access, the exit itself, and the exit discharge.

means of escape   A way out of a building or structure that does not conform to the strict definition of a means of egress, but does provide an alternate way out.

measurement error   The difference between the true value and the value initially obtained by the measuring device.

meatus   A general term for an opening or passageway in the body, such as that of the ear.

meat wrapper's asthma   The respiratory response that may occur among meat packaging personnel as a result of their exposure to contaminants emitted during the cutting and heat-sealing of the polyvinyl chloride plastic wrap used to package meat products.

MEC   Minimum explosive concentration.

mechanical filter respirator   A respiratory protective device that provides protection from airborne particulates, such as dusts, mists, fumes, fibers, and other particulate-type contamination.

mechanical noise   Noise due to impact, friction, or vibration.

mechanical room (Indoor Air Quality)   The area where the air handling equipment, boilers, chillers, and other mechanical equipment for a building is operating.

mechanical ventilation   Air movement caused by a fan or other type of air moving device.

mechanism of action (Toxicology)   The exact molecular process by which a toxicant or one of its metabolites causes an adverse health effect.

MED   Minimal erythemal dose.

media   General term referring to the substance or material on which or in which a contaminant is collected. The media can be a liquid absorbent, solid absorbent, filter, or other material. Typically referred to as the sampling media.

media blank   Clean sampling media that is a new sampler sent to the laboratory with the field samples. Often the laboratory includes media blanks of the same lot number as the field sampling media in the analytical procedure. Used in sample analysis to determine the contribution of the sampling media to the analytical result.

median   The point below which there are as many observations as there are above. The midpoint or middle value of the data or observations.

median effective dose   The dose of a substance required to produce a specific nonlethal effect in 50 percent of a population of the same species.

median lethal concentration   [LC_{50}]   The concentration of a substance in air that is lethal to 50 percent or more of those exposed to it.

median lethal dose   [LD_{50}]   The dose of a material/agent necessary to kill 50 percent of those receiving it.

median particle diameter   The particle size, in micrometers, about which an equal number of particles are smaller or larger in size. See also median particle size.

median particle size   The median size of a particle in a distribution of particles by their size in microns. See also median particle diameter.

mediate   To bring about a settlement by the action of an intermediary.

mediation   The act of intervening between conflicting parties to promote reconciliation, compromise, or settlement.

medical contaminated waste   Materials that contain or have come in contact with objects or substances used in patient diagnosis, care, or treatment.

medical evaluation   The examination by a physician to determine the present health status of an individual, including a review of past history, diagnostic tests, and other measures of health status.

medical examiner   A public official who investigates suspicious, sudden, unnatural death in his/her jurisdiction. This role is provided by a coroner in some jurisdictions.

medical gases   Gases used for medical purposes, such as anesthesia and respiratory therapy.

medical history   A record of a person's past health record, including all hazardous materials he/she was exposed to, and any injuries or illnesses that could affect his/her future health status.

medical monitoring   The periodic evaluation of body functions to determine the health status of an individual.

medical pathology   A disorder or disease.

medical removal protection (OSHA)   [MRP]   Requirement of some OSHA regulations to protect a worker who has been removed from a work assignment that has caused an adverse health effect; this protection prevents additional excessive exposure to the offending agent and provides the worker with income protection and other benefits for a period of time.

medical surveillance program   A medical program that calls for detailed physical examinations for a specific effect or purpose. A periodic review of workers' health status.

medical treatment (OSHA)   Treatment administered by a physician or by registered professional personnel under the standing orders of a physician. Does not include first aid treatment.

medium   A substance through which something is transmitted or carried on, such as air being the medium for transmitting particulates for inhalation.

mega   [M]   Prefix indicating $1 E^6$.

megahertz   [MHz]   One million hertz.

megawatt   [MW]   One million watts of electrical power.

meiosis   A method of cell division.

mel   A 1000 hertz tone, 40 decibels above a listener's threshold, produces a pitch of 1000 mels.

melanin   Dark pigment found in the skin, retina, and hair.

melanocyte   Cell containing dark pigment.

melanoma   A malignant tumor of the pigment-producing cells that contains dark pigment.

melting point   [mp]   The temperature at which a solid changes to the liquid phase.

membrane filter   *See* mixed cellulose ester filter.

Meniere's disease   A combination of deafness, tinnitus, and vertigo with progressive loss of hearing. The disease often develops after a blow to the head or from an infection of the middle ear.

meniscus   The curved surface of a liquid caused by the attraction between a liquid and the walls of a cylinder.

meq   Milliequivalent.

mercantile occupancy (Life Safety)   The occupancy or use of a building, structure, or any part thereof, for the displaying, selling, or buying of goods, wares, or merchandise.

mercurialism   Chronic mercury poisoning due to repeated exposure to elemental mercury or its vapor.

meson   A short-lived unstable particle with or without electric charge that generally weighs less than a proton and more than an electron.

mesothelioma   A relatively rare form of cancer that develops in the lining of the pleura or peritoneum. A malignant tumor of the membrane that surrounds the internal organs of the body.

metabolic pathway   Cellular chemical reactions and transformations that molecules undergo. Biochemical reactions in living cells in which molecules undergo various chemical reactions and transformations.

metabolic rate   The calories (or Btus) required by the body to sustain vital functions, such as the action of the heart and breathing. The rate depends on the physical activity of the individual and physiologic factors.

metabolic reactions   Reactions that include hydrolysis, oxidation, reduction, and conjugation (alkylation, esterification, and acylation).

metabolism   The sum of the physical and chemical processes by which living organized substance is produced and maintained and by which energy is made available for use by the organism.

metabolite   Substance produced by metabolism or by a metabolic process.

metal fume fever   An acute condition caused by exposure to freshly generated fumes of metals, such as zinc, copper, magnesium, and others. An occupational disorder occurring among those engaged in operations in which there is exposure to volatilized metals. Metal fume fever is characterized by malaria-like symptoms and is referred to as the shakes in industries where this occupational disease occurs.

metallizing   The spraying of melted atomized metal onto a surface to be coated.

metamorphosis   A change in the structure and habits of an animal during normal growth, usually in the post-embryonic stage.

metaplasia   The transformation of some types of living cells into a form that is abnormal for the tissue or location in the body.

metastasis   Transmission of disease from an original site to one or more other sites elsewhere in the body. Spread of malignancy from the site of primary cancer to a secondary site by transfer through the lymphatic or blood system. The transfer of disease from one organ or part to another not connected with it. Metastasis may be due to the transfer of pathogenic microorganisms or to the transfer of cells, as occurs in malignant tumors.

metastasize   To be transferred, transmitted, or transformed by metastasis.

meter   The international unit of length equal to 39.37 inches. Also referred to as the metre.

meter   Device used to measure volumes, quantities, or rates of flow of gases, liquids, or electric current.

meter–kilogram–second system   [mks]   A metric system of units which forms the basis of the International System of Units.

methemoglobin   A compound formed with hemoglobin as a result of the oxidation of iron present in hemoglobin from the ferrous to ferric state. This form of hemoglobin does not combine with and transport oxygen to tissues.

methemoglobinemia   The presence of methemoglobin in the blood, resulting in cyanosis.

methenamine pill test   Test of the flammability of carpeting using a methenamine tablet (pill).

Met-L-X   A powder-type fire-extinguishing agent for use with metal fires.

metre   The fundamental unit of length in the metric system. Also referred to as the meter.

metric system   International decimal system of weights and measures based on the meter and kilogram.

metric ton   An amount equal to 1000 kilograms, 0.9842 long tons, and 1.1023 short tons.

meV   Millielectron volt(s), 1 E^{-3} eV.

MeV   Mega electron volts or million electron volts, 1 E^6 eV.

mfpcf   Million fibers per cubic foot (Former method to express airborne asbestos fiber concentration.)

mg   Milligram(s), 1 E^{-3} g.

mG   Milligauss, 1 E^{-3} G.

mg/kg   Milligrams per kilogram.

mg/L   Milligram(s) per liter or parts per million.

mg/m³   Milligram(s) per cubic meter, 1 E^{-3} g/m^3.

mGy   Milligray(s), 1 E^{-3} Gy.

MHz   Megahertz, million cycles per second, 1 E^6 Hz, 1 E^6 cps.

mi   Mile(s).

miasma   Theory that held that all infectious diseases came from putrefying organic matter and resulted in epidemics of cholera, yellow fever, or typhoid.

micatosis   Pneumoconiosis due to inhalation of mica particles.

micelle   A submicroscopic aggregation of molecules.

micro   Prefix indicating one-millionth, 1 E^{-6}.

microbar   A unit of pressure commonly used in acoustics and equal to 1 dyne per square centimeter.

microbe   A microscopic organism.

microbiology   The science that deals with microorganisms, especially their effects on other forms of life.

microcurie   One-millionth of a curie, 1 E^{-6} Ci.

microenvironment   A defined location, such as the home, office, or work area.

microgram   Unit of mass equal to one-millionth of a gram, 1 E^{-6} g.

micromicrocurie   One-trillionth of a curie, 1 E^{-12} Ci.

micro-meteorology   The meteorological characteristics of a local area that is usually small in size (e.g., acres or several square miles) and is often limited to a shallow layer of the atmosphere near the ground.

micrometer   One-millionth of a meter. One micron, 1 E^{-6} m.

micromole   One-millionth of a mole, 1 E^{-6} mol.

micron   One-thousandth of a millimeter, 1 E^{-3} mm, or one-millionth of a meter, 1 E^{-6} m. The unit is used to describe the size of airborne particulate matter.

microorganisms   Minute biological organisms, such as bacteria, molds, viruses, etc.

microphone   An electroacoustic transducer that responds to sound waves and delivers essentially equivalent electric waves. A conduit for producing amplified sound.

microwaves   Non-ionizing electromagnetic radiation in the frequency range from 30 megahertz (MHz) to 300 gigahertz (GHz). If the microwave energy is at a proper wavelength and sufficiently intense, it can damage tissue by heating.

microwave oven   Oven that is designed to heat, cook, or dry food through the application of electromagnetic energy, and that is designed to operate at a frequency of 916 megahertz (MHz) or 2.45 gigahertz (GHz).

microwave radiation   Non-ionizing electromagnetic radiation in the frequency range from 30 megahertz (MHz) to 300 gigahertz (GHz). It can damage living cells by heating them excessively.

middle ear   The region of the ear that contains tiny bones that transfer vibrations from the eardrum to the inner ear.

midget impinger   A sampling device that can be used to collect dusts for concentration determinations by the light field microscopic technique or to collect materials (gases, mists, or vapors) by absorption in a liquid absorbent material. The sampling rate for dusts by this method is 0.1 cubic feet per minute, and the result is expressed as millions of particles per cubic foot of air.

midnight dumping   The disposing of hazardous waste in the darkness or by another illegal manner.

migration (Air Sampling)   The undesired transfer of an adsorbed material from the front section of a solid sorbent tube to the back-up section.

MIG welding   Metal inert gas welding.

mil   Unit of length equal to one-thousandth of an inch, 1 E^{-3} in.

milaria   Skin changes associated with sweat retention.

miliary   Characterized by the formation of blisters or lesions resembling small seeds.

milieu   Environment or surroundings.

milinch   One-thousandth of an inch, 1 E^{-3} in.

milli   Prefix indicating one-thousandth, 1 E^{-3}.

milliamp   [mA]   A unit of electric current equal to one-thousandth of an ampere.

millibar   [mb]   One-thousandth of the standard barometric pressure, 1 E^2 newtons per square meter, or 9.87 E^{-4} bar.

millicurie   One-thousandth of a curie, 1 E^{-3} Ci.

milliequivalent   [meq]   One-thousandth of an equivalent weight of a substance.

milligram   One-thousandth of a gram, 1 E^{-3} g.

milliliter   One-thousandth of a liter, 1 E^{-3} L.

millimeter   One-thousandth of a meter, 1 E^{-3} m.

millimeter of mercury   [mm Hg]   A unit of pressure equal to that exerted by a column of liquid mercury one millimeter high at standard temperature.

millimicron　Unit of length equal to one-thousandth of a micron, 1 E $^{-3}$ micron.

millimole　One-thousandth of a mole, 1 E $^{-3}$ mol.

millions of fibers per cubic foot of air　[mfpcf]　Former unit for expressing the airborne concentration of asbestos fibers in air.

millions of particles per cubic foot of air　[mppcf]　Former unit for expressing the airborne concentrations of dusts, such as coal dust.

millirem　[mrem]　One-thousandth of a rem.

milliroentgen　[mR]　A submultiple of the roentgen, equal to one-thousandth of a roentgen, 1 E $^{-3}$ R.

millisecond　One-thousandth of a second, 1 E $^{-3}$ sec.

millitesla　1 E $^{-3}$ T.

millivolt　[mV]　One-thousandth of a volt, 1 E $^{-3}$ V.

min.　Minimum.

min　Minute(s).

Minamata disease　A neurologic disorder caused by alkyl mercury poisoning, typically characterized by peripheral and circumoral parasthesia, ataxia, dysarthria, and loss of peripheral vision, and leading to permanent neurologic and mental disability or death.

mineral dust　Dust from minerals, such as coal, silica, mica, talc, etc.

mineral spirits　Flammable petroleum product used primarily as a paint solvent and thinner.

mineral wool　A man-made mineral fiber material made from various types of silicate rock. Also referred to as rock wool or slag wool, depending upon the source.

miner's asthma　Asthma associated with anthracosis.

miner's phthisis　Anthracosilicosis.

Mine Safety and Health Administration　[MSHA]　A U.S. federal agency that regulates matters pertaining to health and safety issues regarding mining operations and the mineral industry. It carries out inspections and investigations, enforces regulations, provides technical support, develops relevant training programs, and assesses penalties for violations of regulations.

minimal erythemal dose　Smallest radiant exposure (e.g., UV radiation) that produces a barely perceptible reddening of the skin that disappears after 24 hours.

minimum　The least amount or lowest limit.

minimum audible field (Acoustics)　[MAF]　Identifies the minimum level of simple tones that can just be heard in a very quiet location. They make up the hearing threshold curve.

minimum design duct velocity　*See* transport velocity.

minimum detectable quantity (Instrument)　The amount of material (e.g., micrograms) that gives a response equal to twice the detector noise level.

minimum detectable sensitivity (Instrument)　The smallest amount of input concentration that can be detected as the concentration approaches zero.

minimum detection limit   The lowest concentration or weight of a substance that an instrument can reliably quantify.

minimum lethal dose   [MLD]   The smallest dose that kills one of a group of test animals.

minimum transport velocity   The minimum velocity necessary to transport particulates through a ventilation system without their settling out.

minor   Individual less than 18 years of age.

minute volume   The volume of air inhaled in 1 minute. That is, the product of the tidal volume times the respiratory frequency in 1 minute.

miosis   Contraction of the pupil of the eye.

miotic   An agent that causes the pupil to contract.

miscible   Capable of being mixed in any concentration without separation of phases.

mist   Small droplets of a material that is ordinarily a liquid at normal temperature and pressure.

mite   Any of numerous small arachnids that often infest animals, plants, or stored food.

mitigate   To moderate or reduce the force or intensity of a quality or condition.

mitosis   Nuclear cell division in which the resulting nuclei have the same number and kind of chromosomes as the original cell.

mixture   A heterogeneous association of substances that cannot be represented by a chemical formula.

mjp   Mudded joint packing as insulation for elbows, valves, etc., that may contain asbestos.

mks   Meter–kilogram–second (system).

mL   Milliliter (one-thousandth of a liter), 1 E^{-3} L.

MLD   Minimum lethal dose.

mm   Millimeter(s), 1 E^{-3} m.

mm²   Square millimeters.

mm³   Cubic millimeters, 1 E^{-3} m³.

MMA welding   Manual metal arc welding.

mmad   Mass median aerodynamic diameter, or aerodynamic mass median diameter.

mmcf   Million cubic feet.

mmHg   Millimeter of mercury.

MMMF   Man-made mineral fiber.

mmol   Millimole, 1 E^{-3} mol.

MMVF   Man-made vitreous fiber.

MMWR   Morbidity and Mortality Weekly Report.

mo   Month.

mode   The number that appears most often in a set of data.

model   A mathematical representation of real phenomenon that serves as a pattern from which interrelationships can be identified, analyzed, altered, or synthesized without disturbing the real world situation. A mathematical and/or physical representation of real world phenomena that serves as a plan or pattern from which interrelationships can be identified, analyzed, synthesized, and altered without disturbing real world processes.

modem   Modulator/demodulator. A device employed to transform signals for transmission of information and data by telephone lines. A communication device that allows information to be exchanged between computers via telephone lines.

moderator (Nuclear Industry)   A material, such as beryllium, graphite (carbon), or water, that is capable of reducing the speed of neutrons, thereby increasing the likelihood for them to produce fission in a nuclear reactor.

modulate (Physics)   To vary the frequency, amplitude, phase, or other characteristic of any carrier wave.

modus operandi   The manner in which someone or something operates.

mol   Mole.

mol   Molecule.

molal (Solution)   [m]   A solution containing one mole of solute per 1000 grams of solvent.

molar (Solution)   [M]   A solution containing 1 mole of solute per liter of solution.

molarity   The number of moles of a solute per liter of solution.

molar volume   The volume occupied by a gram mole of a substance in its gaseous state. This is equal to 22.414 liters at standard conditions (temperature of 0°C and 760 millimeters of mercury pressure) and to 24.465 liters at normal temperature and pressure (25°C and 760 millimeters of mercury Hg) in industrial hygiene work.

mold   Any of various fungous growths often causing disintegration of organic matter.

mole   [mol]   Mass of an element or compound equal to its molecular weight. The amount of pure substance containing the same number of chemical units as there are atoms in exactly 12 grams of carbon 12 (i.e., 1 E^{23}). *See* gram-mole or gram molecular weight (mol).

molecule   [mol]   Smallest quantity of a compound that can exist by itself and retain all the properties of the original substance.

molecular weight   [mol. wt.]   The sum of the atomic weights of all the constituent atoms in a molecule.

mole percent   The ratio of the number of moles of one substance to the total number of moles in a mixture of substances, multiplied by 100.

mol. wt.   Molecular weight.

monaural   Indicating sound reception by only one ear.

Monday morning heart attack   Term used to describe heart attacks observed among dynamite workers. The effect is believed to be the result of the vasodilatory effect of ethylene glycol dinitrate and nitroglycerine, which are used in dynamite manufacture.

monel   Term for a large group of corrosion-resistant alloys of predominantly nickel and copper with very small percentages of carbon, manganese, sulfur, and silicon. Some may contain aluminum, titanium, and cobalt.

monitoring   The collection of samples, or the direct measurement of a material concentration/physical agent level, to determine the amount of a substance or physical agent to which a worker is exposed, or which is present in a space, occupied area, or breathing zone of a worker.

monitoring strategy   The plan for implementing and carrying out a monitoring campaign to determine worker exposure to a contaminant, physical agent, etc.

monitoring well   A well used to obtain water samples to test water quality, to measure groundwater levels, or to identify whether a contaminant is present.

mono   Prefix denoting one.

monochromatic   Having only one color, or producing light of only one wavelength.

monodisperse aerosol   A uniform aerosol with a standard deviation of 1.0. That is, the aerosol is all of one size.

monoenergetic (Ionizing Radiation)   Radiation emissions having the same energy.

monomer   A molecule or compound, usually containing carbon and of relatively low molecular weight and simple structure, that is capable of conversion to a polymer, synthetic resin, or elastomer by combination with itself or other similar molecules or compounds.

monthly average (Effluent or Emission)   The arithmetic average of eight sample and analysis data points during any calendar month.

morbid   Diseased.

morbidity   The condition of being sick or morbid. The ratio of sick to well persons in a population.

Morbidity and Mortality Weekly Report   A CDC publication that provides information on current trends in the nation's health.

morbidity rate (Epidemiology)   The rate of disease, injury, or disability from a specific cause per indicated population at a given time.

mordant   A substance that is capable of binding a dye to a textile fiber.

more-than-first-aid injury   An injury for which, in the opinion of the attending physician or nurse, the injured receives more than simple first aid.

moribund   In a condition of impending death and characterized by a deepening stupor or coma.

morphology   Structural configuration. The science of the forms and structure of organized beings and other materials (e.g., objects).

mortality rate   The ratio of the total number of deaths to the total population at a given time.

MOS   Metal oxide semiconductor or metal oxide sensor.

motile   Moving, or having the power to move spontaneously.

motor nerve   A nerve that stimulates muscle contraction.

mottled   Covered with spots or streaks of different shades or colors.

mouth-to-mouth resuscitation   A method of reestablishing respiration by tilting the patient's head back (while the patient is lying on his/her back), pinching the nostrils closed, and exhaling air into the patient's mouth, then removing the mouth to permit the patient to exhale the air forced into the lungs. The procedure is continued until the patient starts to breathe or until it is decided that continuing the procedure is to no avail.

moving curtain air filter   A roll of filter media that extends across the air stream with a motorized system that feeds clean media across the air stream and winds the dirty media onto a take-up spool.

mp   Melting point.

MPC   Maximum permissible concentration.

mph   Miles per hour.

MPI   Mass psychogenic illness.

mps   Meter(s) per second.

mppcf   Million particles per cubic foot of air based on impinger samples counted by the light-field microscopic technique.

mR   Milliroentgen(s), $1 E^{-3}$ R.

mrad   Millirad(s), $1 E^{-3}$ rad.

mrem   Millirem(s), $1 E^{-3}$ rem.

mrem/h   Millirem(s) per hour, $1 E^{-3}$ rem/h.

MRI   Magnetic resonance imaging.

MRT   Mean radiant temperature.

ms   Millisecond.

MS   Mass spectrometer.

MSD   Musculoskeletal disorder.

MSDS   Material safety data sheet.

MSHA   U.S. Mine Safety and Health Administration.

mSv   Millisievert(s), $1 E^{-3}$ Sv.

mT   Millitesla, $1 E^{-3}$ T.

MTD   Maximum tolerated dose.

MUC (Respiratory Protection)   Maximum use concentration.

mucociliary clearance   Removal of materials from the respiratory tract via ciliary action.

mucous membrane   The membrane lining all the channels in the body that communicate with the air, such as the respiratory tract, stomach, urinary tract, intestines, and the alimentary tract, the glands of which secrete mucus.

mucus   The viscous suspension of mucin, water, cells, and inorganic salts secreted as a protective lubricant coating by glands in the mucous membranes.

muffler (Acoustics)   A device for reducing noise emissions from engine exhausts, vents, etc. Two types of mufflers, the dissipative and reactive, are available.

multiple chemical sensitivity   A condition in which an individual is sensitive to a number of chemicals at very low concentrations.

multiple myeloma   A malignant neoplasm of plasma cells usually arising in the bone marrow and manifested by skeletal destruction, pathologic fractures, and bone pain.

musculoskeletal system   Pertaining to or comprising the skeleton and the muscles.

mutagen   A substance capable of causing genetic damage.

mutagenesis   The process in which normal cells are converted into genetically abnormal cells.

mutagenic   An agent that induces genetic mutation.

mutagenicity   The property of being able to induce genetic mutation.

mutant   An organism that possesses a trait that is not inherited, but is the result of a change in its genes.

mutation   Sudden alteration of the hereditary pattern or inheritable characteristic of an individual, animal, or plant. A permanent transmissible change in the genetic material, usually in a single gene.

mutually exclusive events   Events that cannot occur simultaneously.

mV   Millivolt(s), 1 E $^{-3}$ V.

mW   Milliwatt(s), 1 E $^{-3}$ W.

MW   Megawatt(s), 1 E^6 W.

MW   Molecular weight (abbreviation used in some texts).

myalgia   Pain in a muscle or muscles.

mycosis   An infection or disease caused by fungi.

mycotic infection   A fungous infection.

mycotoxin   Any toxic chemical that is produced naturally by a fungus.

mydriasis   Excessive dilation of the pupils.

myelo   Prefix denoting marrow.

myelogenous   Produced in the bone marrow.

myelogenous leukemia   Leukemia arising from myeloid tissue.

myeloid tissue   Tissue pertaining to, derived from, or resembling bone marrow.

myeloma   A tumor composed of cells of the type normally found in the bone marrow.

myelotoxin   A cytotoxin that destroys bone marrow cells.

myelotoxicity   Deterioration of the bone marrow structure that results in dangerous changes in blood composition.

myocardial infarction   Necrosis of the myocardium as a result of an interruption of the blood supply to the area.

myocarditis   Inflammation of the muscular walls of the heart.

myocardium   The middle and thickest layer of the heart wall.

myopathy   A disease or abnormal condition of striated muscle.

myopia   Nearsightedness, a condition in which distant objects appear blurred because their images are focused in front of the retina.

myositis   Inflammation of a voluntary muscle.

N

n   Nano, 1 E $^{-9}$.

N   Newton.

N   Normal (solution).

NAA   Non-attainment area.

NAAQS   National Ambient Air Quality Standards.

NACOSH   National Advisory Committee on Occupational Safety and Health.

nadir   Lowest point.

NAM   National Association of Manufacturers.

NAMS   National Air Monitoring Station.

nano   [n]   Prefix indicating one-billionth, 1 E $^{-9}$.

nanocurie   [nCi]   Unit of radioactivity equal to one-billionth of a curie, 1 E $^{-9}$ Ci.

nanogram   [ng]   One-billionth of a gram, 1 E $^{-9}$ g.

nanometer   [nm]   The billionth part of a meter, 1 E $^{-9}$ m.

nanosecond   [nsec]   A unit of time equal to one-billionth of a second, 1 E $^{-9}$ sec.

narcosis   A reversible stupor or state of unconsciousness that may be produced by some chemical substances.

narcotic (Industrial Hygiene)   A chemical substance that makes a person drowsy.

NAS   National Academy of Sciences.

NASA   National Aeronautics and Space Administration.

nasal septum   The dividing wall between the nasal cavities.

nascent   Coming into existence or in the process of emerging.

nasopharyngitis   Inflammation of the nasopharynx, which is situated above the soft palate at the roof of the mouth.

nasopharynx   The part of the pharynx that lies above the level of the soft palate.

natality   Birth rate.

National Advisory Committee on Occupational Safety and Health   [NACOSH]   Committee established to advise, consult, and make recommendations to the Secretary of Health and Human Services on matters regarding administration of the Department of Labor's Occupational Safety and Health Act.

National Ambient Air Quality Standards   [NAAQS]   Standards in the Clean Air Act (U.S.) that set maximum concentrations for air pollutants to establish nationwide air quality levels that have been judged to be necessary to protect the public health.

National Cancer Institute   [NCI]   An institute that supports research and the dissemination of information related to occupational cancer hazards as well as other causes of cancer.

National Council on Radiation Protection   [NCRP]   An advisory group chartered by the U.S. government to develop and make recommendations on ionizing radiation protection in the U.S.

National Electric Code   [NEC]   Document published by NFPA which provides guidance for the safeguarding of persons, structures, and their contents against hazards arising from the use of electricity. Details are presented in NFPA 70 as well as in the National Electric Code Handbook.

National Electrical Manufacturers' Association   [NEMA]   An association that defines and recommends safety standards for electrical equipment.

National Emission Standards for Hazardous Air Pollutants   [NESHAPS]   Standards set by the EPA under the Clean Air Act for air pollutants that could contribute to an increase in morbidity and mortality.

National Fire Codes   Codes, standards, recommended practices, and manuals developed by the NFPA technical committees.

National Fire Protection Association   [NFPA]   A voluntary, nonprofit association committed to making both the home and the workplace more firesafe. Members promote scientific research into the development and updating of fire safety awareness and produce information and practical publications on fire safety that are of interest to all who are concerned with the preservation of life and property from fire.

National Institute for Occupational Safety and Health   [NIOSH]   That part of the U.S. Department of Health and Human Services that is responsible for investigating the occurrence and causes of occupational diseases and for recommending appropriate standards to the Occupational Safety and Health Administration.

National Institutes of Health   [NIH]   A section of the Public Health Service that conducts research related to diseases and body injuries, and helps establish burn treatment centers.

Nationally Recognized Testing Laboratory   [NRTL]   A laboratory that has been accredited (i.e., for 5 years) by OSHA to test and certify products that require certification under the agency's safety and health standards.

National Occupational Health Research Agenda   A NIOSH coordinating group established to provide guidance on national occupational health and safety research efforts.

National Oceanic and Atmospheric Administration   [NOAA]   An organization within the U.S. Department of Commerce that explores, maps, and charts the oceans and their living resources, and predicts conditions in the oceans, atmosphere, and space.

National Paint and Coatings Association   A national association of paint and coatings manufacturers that developed the HMIS placard/label system for warning about the hazards associated with the use of their products, as well as for wider application.

National Pollutant Discharge Elimination System   [NPDES]   A national program for issuing, modifying, revoking, reissuing, terminating, monitoring, and enforcing discharge permits, and for imposing pretreatment requirements under the EPA Clean Water Act.

National Research Council   [NRC]   A council established by the National Academy of Sciences that serves to provide scientific and technical advice to the government, the public, and the scientific and engineering communities.

National Response Center   [NRC]   Contact for reporting spills of dangerous materials.

National Safety Council [NSC] Independent nonprofit organization that provides information, literature, training, and support for occupational safety- and health-related programs and issues with the goal of reducing the number and severity of accidents/occupational diseases in the U.S. and preventing their occurrence.

National Technical Information Service [NTIS] A federal agency that develops and provides technical information on products and services.

National Toxicology Program [NTP] A program established to determine the toxic effect of chemicals and to develop more effective and less expensive toxicity test methods.

National Water Quality Standards Minimum requirements for water quality that require states to set standards to achieve the Clean Water Act's fishable and swimmable goal for water quality.

natural draft The negative pressure created by the height of a stack or chimney and the temperature difference between the flue gas and the outside.

natural frequency (Vibration) The frequency at which an undamped system will oscillate when momentarily displaced from its rest position.

natural gas A combustible gas composed chiefly of methane and ethane, as well as smaller amounts of other combustible gases.

natural hazard The potential for events in nature, such as a hurricane, earthquake, tornado, or other natural occurrence, to threaten the health and welfare of the people.

natural immunity Immunity that is present at birth and that is not artificially acquired.

naturally occurring radioactive material [NORM] Any nuclide that is radioactive in its natural physical state, but does not include source material or special nuclear material (i.e., plutonium, uranium 233, or uranium enriched in the isotopes uranium 233 or uranium 235).

natural radioactivity The property of radioactivity exhibited by more than fifty naturally occurring radionuclides.

natural uranium Uranium as found in nature.

natural ventilation Air movement created by wind, a temperature difference, or other non-mechanical means. The movement of outdoor air into a space through intentionally provided openings, such as doors or windows, as well as by infiltration.

natural wet-bulb thermometer A thermometer whose temperature is indicated by a wetted thermometer bulb that is exposed to and cooled by the movement of the surrounding air.

nausea An unpleasant sensation, often culminating in vomiting.

nauseous Experiencing symptoms of nausea.

nautical mile A unit of length equal to 6076 feet or 1852 meters.

NBS National Bureau of Standards.

nCi Nanocurie(s).

NCI National Cancer Institute.

NCRP National Council on Radiation Protection and Measurements.

NDIR Nondispersive infrared.

near field (Acoustics)   The area close to a sound source within which the sound pressure level does not obey the inverse square law concept (i.e., reduction of 6 dBA for each doubling of distance from the noise source).

near field (Electromagnetic Radiation)   Region near a radiating electromagnetic source or structure in which the electric and magnetic fields do not have a substantially plane wave character, but vary considerably from point to point. Typically, the near field extends out to at least 5 wavelengths from the radiating device.

near miss   An accident having significant potential to cause damage to equipment, personal injury, or other form of harm, but where no damage, injury, or harm occurred.

nebulizer   A device used to provide a fine mist of a liquid.

NEC   National Electrical Code.

necropsy   Examination of the body after death.

necrosis   Destruction of body tissue. Tissue death.

necrotic   Characterized by necrosis or destruction.

needle stick   A puncture of the skin by a needle that may have been contaminated by contact with an infected patient.

needle valve   Valve that has a sharp-pointed needle-like plug that can be moved into and out of a cone-shaped seat to control flow.

negate   To nullify or render ineffective and invalid.

negative pressure (Ventilation)   Condition that exists when less air is supplied to a space than is exhausted from the space, so the air pressure within that space is less than that in the surrounding area.

negative pressure enclosure (Asbestos)   An enclosed area that is provided with four air changes per hour and is maintained at a negative 0.02 inches of water pressure differential relative to the pressure outside the work area.

negative pressure respirator   A respirator in which the pressure inside the respirator is negative during inspiration relative to the pressure outside, and positive during exhalation relative to the pressure outside.

negligence   Failure to do what reasonable and prudent persons would do or would have done under similar circumstances.

NEL   No-effect level.

NEMA   National Electrical Manufacturers' Association.

nematode   Worms having threadlike bodies, many of which are parasitic to man.

neonatal   Pertaining to the first four weeks after birth.

neonate   A newborn child.

neoplasia   The progressive multiplication of cells under conditions that would not elicit the multiplication of normal cells.

neoplasm   Any new and abnormal growth, such as a tumor. A new growth of tissue in which the growth is uncontrolled and progressive.

nephelometer   An instrument which measures the scattering of light due to particles suspended in a medium (e.g., water).

nephelometry   Photometric analytical technique for measuring the light scattered by finely divided particles of a substance in suspension.

nephritis   Inflammation of the kidneys.

nephrology   The scientific study of the kidney, its anatomy, physiology, and pathology.

nephropathy   Any disease of the kidney.

nephrosis   Any disease of the kidney.

nephrotoxic agent   *See* nephrotoxin.

nephrotoxin   A substance that is harmful to the kidney.

nepotism   Favoritism shown to close friends or relatives.

nerve   A bundle of fibers interconnecting the central nervous system and the organs or parts of the body that are capable of transmitting sensory stimuli and motor impulses from one part of the body to another.

nervous system   A coordinating mechanism that regulates internal body functions and the response to external stimuli.

NESHAPS   National Emission Standards for Hazardous Air Pollutants.

net instrument response   The gross instrument response for the sample, minus the sample blank.

neural   Pertaining to a nerve, the nerves, or the nervous system.

neuralgia   Sudden pain along a nerve.

neural loss (Acoustics)   Hearing loss due to nerve damage.

neurasthenia   Neuroses marked by lack of energy, depression, loss of appetite, insomnia, and inability to concentrate.

neuritis   Inflammation of a nerve.

neurology   The biological science that deals with the nervous system and its behavior.

neuron   A nerve cell that serves as the functional unit of the nervous system.

neuropathy   A general term denoting functional disturbances and/or pathological changes in the peripheral nervous system.

neurosis   Functional disorder of the mind or emotions, without obvious organic lesion or change, and involving anxiety, phobia, or other abnormal behavioral symptoms.

neurotoxic agent   *See* neurotoxin.

neurotoxicology   The study of the effects of toxins on nerve tissue.

neurotoxin   A substance that is poisonous or destructive to nerve tissue.

neutrino   A particle, resulting from a nuclear reaction, that carries energy away from the system, but has no mass or charge.

neutron   Fundamental particle of matter having a mass of 1.009, but no electric charge. One of three basic atomic particles.

neutron capture (Health Physics)   A process in which a nucleus absorbs a neutron.

neutron radiation   Ionizing radiation resulting from nuclear disintegration.

newton   [N]   A unit of force which, when applied to a mass of 1 kilogram will give it an acceleration of 1 meter per second.

NFPA   National Fire Protection Association.

NFPA 704   A method for visually presenting information on the health (green square), flammability (red square), and reactivity or stability (yellow square) of a material, as well as providing special information associated with these hazards for firefighters and others. A fourth square (white) is used to designate water reactivity. Numerals between 0 (no hazard) and 4 (very hazardous) are placed in the three upper squares of the diamond to indicate the degree of hazard of the material as relates to health, flammability, and reactivity.

ng   Nanogram, $1 E^{-9}$ g.

NIBS   National Institute of Business Sciences.

nickel itch   A type of dermatitis seen in some workers who are exposed to nickel.

NIEHS   National Institute of Environmental Health Sciences

night blindness   A lack of normal night vision that is often the result of a vitamin A deficiency.

NIH   National Institutes of Health.

NIHL   Noise-induced hearing loss.

NIOSH   National Institute for Occupational Safety and Health.

NIPTS   Noise-induced permanent threshold shift.

NIST   National Institute of Standards and Technology (formerly the National Bureau of Standards).

nitrogen cycle   The continuous cyclic progression of chemical reactions in which atmospheric nitrogen is compounded, dissolved in rain, deposited in the soil, assimilated and metabolized by bacteria and plants, and returned to the atmosphere by organic decomposition.

nitrogen fixation   The utilization of atmospheric nitrogen to form chemical compounds. In nature this is accomplished by bacteria, and results in the ability of plants to synthesize proteins.

nm   Nanometer(s), $1 E^{-9}$ m.

N/m²   Newtons per square meter.

NMR   Nuclear magnetic resonance. Now known as magnetic resonance imaging, or MRI.

NNI   Noise and number index.

NOAA   National Oceanic and Atmospheric Administration.

NOAEL   No-observed-adverse-effect level.

noble gases   Elements that are completely unreactive or react only to a limited extent with other elements. There are six inert gases that are referred to as noble, or inert gases, including argon, helium, krypton, neon, radon, and xenon.

NOC   Not otherwise classified.

nocturnal   Occurring at night.

node (Acoustics)   A point or region of minimum or zero amplitude in a periodic system.

node (Physiology)   A small mass of tissue that may be normal or pathological.

nodule   A localized swelling or mass of tissue that can be detected by touch. A small node.

NOEL   No-observed-effect level.

NOI   Notice of intent.

noise (Acoustics)   Any undesired sound or unwanted disturbance within the audible frequency range. Excessive or unwanted sound that potentially results in annoyance and/or hearing loss.

noise (Instrument)   Any unwanted electrical disturbance or spurious signal that modifies the transmission, measurement, or recording of desired data. An output signal of an instrument that does not represent the variable being measured, or the variation in the signal from an instrument that is not caused by variations in the concentration of the material being measured.

noise and number index   [NNI]   An index used for rating the noise environment near airports and the noise associated with aircraft flybys.

noise contour   A continuous line on a plot plan or map which connects all points of a specified noise level, (e.g., 85 dBA).

noise-induced hearing loss   [NIHL]   Progressive hearing loss that is the result of exposure to noise, generally of the continuous type, over a long period of time, as opposed to acoustic trauma that results in immediate hearing loss.

noise level (Acoustics)   For airborne sound, unless specified to the contrary, the weighted sound pressure level, called sound level, the weighting of which must be indicated (e.g., A, B, or C weighting). *See* sound level.

noise reduction   The reduction in the sound pressure level of a noise, or the attenuation of unwanted sound by any means.

noise reduction rating   [NRR]   An empirical technique developed by the EPA for determining and indicating the noise-attenuating capability of a hearing protective device.

nolo contendere   Literally, a legal plea that one does not wish to contend or contest an allegation. Often taken as an admission of guilt in a legal proceeding.

nomograph   A chart in the form of linear scales that represent an equation containing a number of variables so that a straight line can be placed across them, cutting the scales at values of the variables, satisfying the equation, and yielding an answer to that which one is solving for.

non-asbestiform fiber   A fibrous material which contains no asbestos.

non-attainment area   [NAA]   Regions or areas of the country that do not meet the national ambient air quality standards of the Clean Air Act for one or more of the federally regulated air pollutants.

non-auditory effects of noise   Effects from exposure to noise, such as stress, fatigue, reduction in work efficiency, etc.

noncompliance   A situation in which there is a failure to follow and adhere to the rules and regulations or to the advice provided.

nondestructive testing   The examination of an item to identify defects without damaging it.

nondisabling injury   An injury for which the injured person loses less than one full shift from work. Includes a first aid injury, or more-than-first-aid injury and a restricted duty injury.

nondispersive infrared   [NDIR]   A measurement principle that can be employed to measure the airborne concentration of some materials (e.g., CO, CO_2, etc.) using an infrared source and photocell to determine the absorption of the infrared radiation, and that is dependent on contaminant concentration in the sample.

nonflammable   A material or substance that will not burn readily or quickly.

nonflammable gases   Gases that will not burn in any concentration of air or oxygen.

nonfriable (Asbestos)   A material that contains more than 1 percent asbestos (by weight) and that cannot be crumbled by hand pressure when dry.

non-intact skin   Skin that has eruptions, is weeping, has a rash, or is chapped or abraded.

non-ionizing radiation   Any form of radiation that does not have the capability of ionizing the medium through which it is passing.

nonliquefied compressed gas   A gas, other than a gas in solution, which, under the charged pressure, is entirely gaseous at a temperature of 70°F.

nonparametric (Statistics)   Statistical methods that do not assume a particular distribution for the population under consideration.

nonpareil   Without equal.

nonpermit confined space (OSHA)   A confined space that does not contain or, with respect to atmospheric hazards, does not have the potential to contain any hazard capable of causing death or serious physical harm.

nonpolar compound   A compound for which the positive and negative electrical charges coincide and for which the molecules do not ionize in solution and impart electrical conductivity.

nonpolar solvents   The aromatic and petroleum hydrocarbon group of compounds are considered nonpolar solvents.

nonrandom sample   Any sample taken in such a manner that some members of the defined population are more likely to be sampled than others.

nonroutine respirator use   The wearing of a respirator when carrying out a special task that occurs infrequently.

nonsequitur   An inference or conclusion that does not follow from the evidence.

nonserious violation citation (OSHA)   Citation issued when a situation would affect worker safety or health, but would not cause death or serious physical harm.

nonspecific effect   An effect that is not caused by a specific agent or contaminant.

nonthreshold toxicant   A substance that is known or assumed to produce some risk of adverse effect at all doses above zero.

nontoxic   A material for which experience and/or experiments have failed to cause physiologic, morphologic, or functional changes that adversely affect the health of man or animal.

nonviable   Not capable of living or developing.

nonviable particles Particulates resulting from crushing, grinding, combustion, air impacting on settled dust, or roadway dust, copier paper fines, etc. that contribute to indoor air pollution.

nonvolatile Material that does not evaporate at ordinary temperature.

no-observed-adverse-effect level [NOAEL] The highest dose of a chemical that, in an experiment, caused no significant signs of toxicity.

no-observed-effect level [NOEL] The highest dose of a chemical in an experiment that did not produce a toxic effect.

NORA National Occupational Research Agenda.

norm That which is normal.

NORM (Ionizing Radiation) Naturally occurring radioactive material.

norm (Statistics) A statistical measure of average observed performance.

normal (Solution) [N] A solution containing 1 gram equivalent weight per liter of solution.

normal distribution A symmetrical, bell-shaped distribution of data in which the mean, median, and mode are the same. This distribution is characterized by a maximum number of occurrences at the center or mean point, a progressive decrease in the frequency of occurrences with distance from the center, and a symmetry of distribution on either side of the center. Also called the Gaussian distribution. This type of distribution occurs most commonly in nature and industry.

normal humidity A range of 40 to 80 percent relative humidity.

normal range (Biological Testing) The range of values of a biological analyte that would be expected without exposure to an environmental contaminant in the workplace.

normal solution [N] A solution containing 1 gram-equivalent weight of solute in 1 liter of solution.

normal temperature and pressure (Industrial Hygiene) Normal conditions in industrial hygiene are 25°C and 760 millimeter pressure.

N.O.S. Not otherwise specified.

nosoacusis Hearing loss attributable to heredity, disease (mumps, measles, etc.), drugs, chemicals, barotrauma, and blows to the head.

nosocomial disease A disease with its source in a hospital or other health care facility and that is contracted by an individual as a result of being there.

nosography The systematization and description of diseases.

nosology The branch of medicine that deals with the classification of diseases.

Notice of Proposed and Rule-Making [NPRM] A document issued by a federal agency published in the Federal Register that solicits public comment on a proposed regulatory action.

notifiable disease A disease that, by statuatory requirements, must be reported to the public health authority in the jurisdiction where and when the diagnosis was made.

not otherwise classified [NOC] A category of items, including relatively infrequent dissimilar items.

nourishment   Substance that nourishes and supports life.

NO$_x$ Oxides of nitrogen.

noxious   Injurious or harmful to health.

NOYS   A unit used in the calculation of perceived noise level.

NPA   National Particleboard Association.

NPCA   National Paint and Coatings Association.

NPDES   National Pollutant Discharge Elimination System.

NPE   Negative pressure enclosure.

NPRM   Notice of Proposed Rule-Making.

NPT   National pipe thread.

NRC   National Research Council.

NRC   National Response Center.

NRC   Nuclear Regulatory Commission.

NRDC   Natural Resources Defense Council.

NRR   Noise reduction rating.

NRTL   Nationally Recognized Testing Laboratory.

NSC   National Safety Council.

NSF   National Science Foundation.

NSPS   New source performance standard.

NTIS   National Technical Information Service.

NTP   National Toxicology Program.

NTP   Normal temperature and pressure.

nuclear energy   The energy released as the result of a nuclear reaction. The processes of fission or fusion are employed to create a nuclear reaction.

nuclear fision   A nuclear transformation in which a nucleus is divided into at least two nuclei, with the release of energy.

nuclear reaction   A reaction that alters the energy, composition, or structure of an atomic nucleus.

nuclear reactor   A device in which a chain reaction is initiated and controlled, with the consequent production of heat. Typically utilized for power generation.

Nuclear Regulatory Commission   [NRC]   A federal agency that regulates all commercial uses of nuclear energy, including the construction and operation of nuclear power plants, nuclear fuel reprocessing, research applications of radioactive materials, etc.

nucleon   Common term for the constituent particles of the nucleus, such as the proton or neutron.

nucleus (Biology)   The central portion of a living cell, consisting primarily of nucleoplasm in which chromatin is dispersed.

nucleus (Chemistry)   The positively charged central mass of an atom, containing essentially all the mass in the form of protons and neutrons.

nuclide   A species of atom characterized by the number of protons and neutrons, and by the energy content.

nuisance   A source of inconvenience or annoyance.

nuisance dust   Airborne particulates that do not alter the architecture of the air spaces of the lungs, do not produce scar tissue to a significant extent, and whose tissue reaction is reversible. They are not recognized as the direct cause of a serious pathological condition. *See* inert dust.

null hypothesis   An hypothesis about a population parameter to be tested. Assumes there is a difference between the groups in a study.

numb   Insensible and unfeeling.

NVLAP   National Voluntary Laboratory Accreditation Program (Part of the National Institute of Standards and Technology [NIST].

NWQS   National Water Quality Standards.

nystagmus   Rapid movement of an eyeball.

OA   Outdoor air.

O and M program   Operation and maintenance program.

OBA   Octave band analyzer.

objective (Epidemiology)   The stated end to which efforts are directed, specifying the population, outcome, and variable to be measured.

objective (General)   What is intended to be accomplished.

obligate parasites   Organisms which can only survive in living cells.

observation time (Toxicology)   The time from the first and/or last exposure in an experimental study.

occluded   Closed, shut, or blocked.

occlusion (Hearing)   The reduction of transmission of sound to the inner ear as a result of blockage in the ear canal.

occupancy   The purpose for which a building or portion thereof is used or is intended to be used.

occupant load   The total number of persons that may occupy a building or portion thereof at any time.

occupational biomechanics   The application of biomechanics to the study of the human body in motion and at rest, focusing on the physical interaction of workers with their tools, machines, and materials, in order to optimize human performance and minimize the risk of musculoskeletal disorders.

occupational disease   A disease that is a result of exposure to a hazardous material, physical agent, biological organism, or ergonomic stress in the course of one's work.

occupational dose (Ionizing Radiation)   The dose received by an individual in a restricted area or in the course of employment in which the individual's assigned duties involve exposure to radiation or to radioactive materials from licensed and unlicensed sources, whether the individual is in possession of the license or another person is.

occupational exposure   Exposure to a health hazard, such as a chemical, physical, or biologic agent, or an ergonomic factor, while carrying out work within a workplace.

occupational exposure limit   [OEL]   The concentration of an airborne contaminant or physical stress that is an acceptable level of exposure for a specified period of time.

occupational illness   Any abnormal physical condition or disorder, other than one resulting from an occupational injury, caused by exposure to environmental factors associated with employment. Includes acute and chronic illnesses or diseases, that may be caused by the inhalation, absorption, or ingestion of, or direct contact with, a hazardous substance, or by physical stress or an ergonomic factor.

occupational injury   Any injury, such as a cut, fracture, sprain, amputation, etc., that results from a work accident or from an exposure involving a single incident in the work environment.

Occupational Safety and Health Act   A law whose purpose is to assure as far as possible that every working person in the U.S. and its possessions will have safe and healthful working conditions, thereby preserving our human resources.

Occupational Safety and Health Administration   [OSHA]   A federal agency within the U.S. Department of Labor responsible for establishing and enforcing standards regarding the exposure of workers to safety hazards or harmful materials that they may encounter in the work environment, as well as other matters that may affect the safety and health of workers.

Occupational Safety and Health Review Commission   [OSHRC]   A commission independent of OSHA that has been established to review and rule on contested OSHA cases.

occupied zone (Indoor Air Quality)   Locations/positions where the people work or occupy space within a building. Considered to be the space between planes at 3 and 72 inches (7.5 to 180 centimeters) above the floor and more than 2 feet (30 centimeters) from walls.

occurrence (Epidemiology)   General term describing the frequency of a disease or event in a population without distinguishing between incidence or prevalence.

octane number   A numerical rating used to grade the relative anti-knock properties of gasolines. A high octane fuel (e.g., octane rating of 89 or more) has better anti-knock properties than one with low octane.

oceanography   The science of the ocean and its physical and chemical characteristics.

octave   The interval between two sounds having a frequency ratio of 2 : 1.

octave band   The separation of noise energy into frequency bands that cover a 2 to 1 range of frequencies. The center frequencies of these bands are 31.5, 63, 125, 250, 500, 1000, 2000, 4000, 8000, and 16000 hertz. This separation is used to analyze noise. One-third-octave band and one-tenth-octave band analyses are also used to obtain a more detailed analysis of noise.

octave band analyzer   [OBA]   A portable instrument used for characterizing the frequency and amplitude characteristics of a sound.

ocular   Pertaining to the eye.

o.d.   Outside diameter.

OD   Optical density.

ODC   Ozone-depleting compound.

odds ratio   The ratio of the probability of an event occurring to that of the event not occurring.

odor   The property of a substance that affects the sense of smell.

odorant (Liquefied Petroleum Gas)   An approved agent that effectively odorizes all liquefied petroleum gases that is of such character as to indicate positively, by distinct odor, the presence of gas down to a concentration in air of not over one-fifth the lower limit of flammability.

odorizing gas   An odorant added to liquefied petroleum gases to indicate, by distinct odor, the presence of the gas down to a concentration of not over one-fifth the lower limit of flammability.

odor panel   A group of individuals trained to identify and rate odors in order to check quality and acceptability.

odor threshold   The minimum concentration of a compound in the air that is detectable by odor.

ODTS   Organic dust toxic syndrome.

OEL   Occupational exposure limit.

offal   The intestines and discarded parts of an animal after it has been slaughtered.

off-hours   Typically considered to be from the period between 6:00 p.m. and 6:30 a.m. on normal workdays, and all day Saturday, Sunday, and legal holidays.

office occupancy   The occupancy or use of a building or structure or any portion thereof for the transaction of business or the rendering or receiving of professional services.

Office of Pollution Prevention and Toxics   A section of the U.S. Environmental Protection Agency.

office safety   A safety program for individuals working in an office environment. It recognizes the many hazards to which personnel are exposed, including slips, trips, cuts, falls from standing on chairs or other office furniture, being struck by falling objects, paper cuts, etc. The program identifies hazards and presents appropriate information to reduce the likelihood of injury from poor layout and office design and trip hazards, and to ensure safe electrical practices, effective fire and emergency procedures, satisfactory egress routes, adequate illumination and ventilation, proper storage facilities, etc.

off-shore breeze   The movement of air from the land to the sea that typically occurs after the sun goes down and the land mass cools.

off-the-job safety   That part of the employer's safety program that is directed to employees when they are not at work. The objective of this effort is to get employees to follow the same safe practices in their outside activities as they adhere to on the job.

OGC   Office of General Council.

OH&S   Occupational health and safety.

ohm   The unit of electrical resistance. It is equal to the resistance through which a current of 1 ampere will flow when there is a potential difference of 1 volt across it.

Ohm's law   A law applied to the flow of electricity through a conductor that states that the current flow in amperes is proportional to the voltage divided by the resistance in ohms.

oil dermatitis   *See* oil folliculitis.

oil folliculitis   Acne-like lesions resulting from repeated skin contact with some oil products, such as insoluble cutting oils.

oilless compressor   An air compressor that is not lubricated with oil. Also referred to as a breathing air compressor. Thus, it does not generate carbon monoxide or oil mist when in operation.

oil mist   Aerosol produced when oil is forced through a small orifice, splashed or spun into the air during an operation, or vaporized and then condensed in the atmosphere.

oil patch   The oil field.

oil shale   A sedimentary rock that contains kerogen and that can be processed to yield oil and gas products. The oil and gas is obtained from the shale when the rock is heated in an inert atmosphere to convert the kerogen content of the shale to oil vapors, hydrocarbon gases, and a carbonaceous residue.

oldest air (Indoor Air Quality)   Found anywhere in the indoor air environment, depending on the stagnant air zones in the room.

OLF   A perceived air quality term that attempts to quantify the level of odorous pollutants in OLFS.

olfactory   Pertaining to the sense of smell.

olfactory fatigue   Condition in which the sense of smell has been diminished to the extent that an odor cannot be detected.

oliguria   The secretion of a diminished amount of urine in relation to the fluid intake.

OMB   Office of Management and Budget.

omnidirectional (Acoustics)   Noise incident on a receptor that is perceived to be from all directions for low frequency sounds, but not so for high frequency sounds. Microphones are omnidirectional for low frequency sounds, but become directional as the frequency increases.

oncogenes   Hypothetical viral genetic material carrying the potential of cancer and passed from parent to offspring.

oncogenesis   The production or causation of tumors.

oncogenic   The property of a substance or mixture of substances to produce or induce benign or malignant tumor formations in living animals.

oncogenicity   The quality or property of being able to cause tumor formation.

oncology   The study of tumors, including the study of causes, development, characteristics, and the treatment of tumors.

one-tenth-octave band   A bandwidth equal to one-tenth of an octave. *See* octave band.

one-third-octave band   A bandwidth equal to one-third of an octave *See* octave band.

on-shore breeze   The wind direction during the daytime when the sun warms the land, causing air to rise and, as a result, airflow toward land from the sea. Also referred to as a sea breeze, but this term is not limited to times when the airflow in the on-shore direction is a result of the sun warming the land.

on-stream   In operation or running.

oocyte   Developing egg cell.

opacity   The quality of being opaque or dull.

opacity (Plume)   The degree to which light is reduced, or the degree to which visibility of a background is reduced.

opaque   Impenetrable by light.

OPEC   Organization of Petroleum Exporting Countries.

open burning   Uncontrolled burning in an open area, in outdoor incinerators, or in an open dump, either intentionally or accidentally.

open-circuit SCBA   A type of self-contained respiratory protection device that exhausts exhaled air to the atmosphere rather than recirculating it.

open-cup test   A method employed to determine the flash point of a flammable liquid.

open-face sampling   The collection of airborne particulates employing a filter cassette with the front closure removed.

open-path detectors   Line of sight contaminant detection systems that can cover a wide area. Detection is dependent on the contaminant crossing or breaking the detector line-of-sight measurement beam, such as an infrared or ultraviolet source. Results are typically expressed in ppm-meters.

open system   A system in which the handling or transfer of a material occurs in a manner such that there is contact of the material with the atmosphere.

operation and maintenance program   [O and M program]   A program in which specific procedures and work practices are defined and adhered to in order not to disturb a hazardous material, such as asbestos or lead, thereby reducing the potential for exposure to the substance of concern. The program also includes periodic inspection of the condition of the hazardous material to determine whether or not there is a need for an abatement action (e.g., removal, enclosure, etc.).

OPIM   Other potentially infectious materials.

opportunistic infection   An infection that would normally not affect an individual with a healthy immune system, but that can affect an individual whose immune system has been suppressed, as occurs in persons with AIDS.

OPPT   Office of Pollution Prevention and Toxics.

opthalmologist   A physician who specializes in the structure, function, and diseases of the eye.

ophthalmoscope   An instrument used to examine the interior of the eye.

optical cavity (Laser)   A system of using mirrors to pass a light beam through a lasing medium several times, thereby amplifying the number of photons emitted.

optical density   [OD]   A measure of the total luminous transmittance of an optical material. A logarithmic expression of the attenuation provided by a filter. The logarithmic value of the ratio between the intensity of transmitted light through a clean filter and a sample.

optical fiber   A filament of optical material (e.g., quartz) that is capable of transmitting light along its length by multiple internal reflections and emitting the light at the end of the fiber.

optical microscope   A magnifying lens system utilizing visible light for viewing objects.

optic nerve   A nerve leading from the eye to the brain.

optimal   The most favorable or ideal.

optimum   The best or most favorable condition for a particular situation.

oral   Pertaining to the mouth.

oral ingestion   The swallowing of a material.

oral route   The route of entry into the body through the mouth.

oral toxicity (Toxicology)   The toxic effect as determined by dosing by mouth.

order of magnitude   An estimate of size or magnitude expressed as a power of ten. For example, 1000 is two orders of magnitude more than 10.

ore   A mineral from which a valuable substance such as a metal can be extracted.

organ   An organized collection of tissues that have a specific function.

organelle   A specialized part of a cell that resembles and functions as an organ.

organic chemistry   The chemistry of carbon compounds.

organic disease   A disease in which there is some change in the structure of body tissue.

organic dust toxic syndrome   A toxic effect as a result of exposure to organic dust that may contain plant particles as well as bacterial and fungal products. Symptoms include breathing difficulty, chills, fever, and pneumonia-like symptoms.

organic refuse   Solid waste composed of carbonaceous materials, such as wood, paper, etc.

organic phosphates   *See* organophosphates.

organic solvents   Organic compounds that are typically used as paint thinners and solvents, cleaning agents, and solvents for adhesives.

organism   Any individual living thing, whether animal or plant.

Organization of Petroleum Exporting Countries   [OPEC]   A group of oil-producing nations that have organized to advance member interests in dealing with oil-consuming nations.

organ of Corti   An organ, laying against the basilar membrane in the cochlear duct of the ear, that contains special sensory receptors for hearing.

organogenesis   The period in the development of a fetus during which the organs are developing.

organometallic compound   A chemical compound in which a metal is chemically bonded to an organic compound. Examples include organo-phosphate compounds, tetraethyl lead, manganese cyclopentadienyl tricarbonyl, etc.

organophosphates   A group of chemical pesticides that contain phosphorous.

organophosphorous poisoning   The toxicity resulting from exposure to an organophosphorous chemical. The effect is on the nervous system as a result of the inhibition of acetylcholinesterase. This can result in death due to bronchoconstriction and paralysis of the respiratory centers of the brain stem.

orifice   An opening or hole of controlled size that can be used for the measurement of liquid or gas flow.

orifice meter   A device for determining flow rate. A flowmeter employing, as the measure of flow, the pressure difference as measured on the upstream and downstream side of a specific type restriction within a pipe or duct.

ORM   OSHA reference (Sampling and Analytical) method.

ornithosis   A disease of birds and domestic fowl that is transmissible to man. In man, the disease is called psittacosis.

ORNL   Oak Ridge National Laboratory.

Orsat apparatus   A device for measuring the percentage of carbon dioxide, oxygen, and carbon monoxide in flue gas.

oscillate   To swing back and forth with a steady, uninterrupted motion.

oscillation (Acoustics)   Variation, usually with time, of the magnitude of a quantity with respect to a specified reference when the magnitude is alternately greater and smaller than the reference. The back and forth variation of a steady, uninterrupted sound.

oscilloscope   An instrument that visually displays the shape of electrical waves on a fluorescent screen.

OSHA   Occupational Safety and Health Act/Occupational Safety and Health Administration.

OSHA Reference Method (Asbestos)   [ORM]   The analytical method for collecting and analyzing asbestos samples as spelled out in Appendix A of the OSHA asbestos standard. Analysis must be carried out by this procedure or an equivalent method.

OSHRC   Occupational Safety and Health Review Commission.

osmosis   The passage of fluid through a semipermeable membrane as a result of the difference in concentration on either side of the membrane. The diffusion of a fluid through a semipermeable membrane into a more concentrated solution.

ossicle   A small bone. The auditory ossicles are the malleus, incus, and stapes of the middle ear.

ossification   Formation of bone substance.

osteo   Prefix denoting a relationship to a bone or to the bones.

osteoma   A tumor composed of bone tissue that typically develops on a bone, but can develop on other tissue.

osteomalacia   Softening of the bones.

osteoporosis   Increased porosity of bone.

osteosclerosis   The hardening or abnormal density of bones.

other potentially infectious material   [OPIM]   Body fluids such as saliva, amniotic fluids, or pleural, or any body fluid contaminated with blood.

other-than-serious violation (OSHA)   A violation generally issued if a violation of an occupational safety and health regulation does exist, but the most likely result of the condition would not result in death or serious physical harm.

otitis media   An inflammation of the middle ear.

otolaryngologist   Physician concerned with diseases of the ear, nose, and throat.

otologic   Pertaining to otology.

otologist   A physician who specializes in surgery and diseases of the ear.

otology   Branch of medicine that deals with the ear, its anatomy, physiology, and pathology.

otosclerosis   The formation of bony tissue in the cavities of the inner ear.

otoscope   An instrument for examining or auscultating (examining by listening) the ear.

ototoxic   Medication and other substances that can affect hearing acuity; that is, they are poisonous to the organs of hearing.

ounce (Liquid Measure)   Equal to 29.573 milliliters.

ounce (Weight Measure)   Equal to 28.35 grams.

outage   The space left in a container to allow for expansion due to changes in temperature.

outbreak (Epidemiology)   An epidemic limited to a localized increase in the incidence of a disease.

outdoor air   The fresh air that is brought into a building from the outside and that is combined with return air to supply respirable air to occupied areas.

outer ear   The ear canal that connects the outside with the ear drum.

outfall   The location at which an effluent is discharged into a receiving body of water.

outlet damper (Ventilation)   A damper on a fan outlet of a ventilation system that adds resistance and provides more even flow.

outlier   Observations or results that differ widely from the rest of the data so as to lead one to suspect that a gross error has occurred or that the value(s) came from a different population.

outside stairs   Stairs in which at least one side is open to the outer air.

overall noise   [OA sound pressure level]   The overall sound pressure level in decibels, as determined by a microphone and meter, without the weighting of the frequency components of the noise to attenuate part of the sound spectrum.

overburden   Material overlaying a mineral deposit.

overexposure   With regard to occupational exposure, an exposure to an airborne contaminant or physical stress at a level that is above an established limit, such as an OSHA PEL, and ACGIH TLV, or another recognized exposure limit.

oxidant (Air Pollution)   A substance containing oxygen that reacts chemically in air to produce new substances. These are a source of photochemical smog.

oxidant (Chemistry)   The electron acceptor in an oxidation–reduction reaction.

oxidant (Gas)   Gases that do not burn, but that support combustion.

oxidation (Chemistry)   Any reaction in which electrons are transferred from one substance to another.

oxidation pond   A holding area where organic wastes are decomposed by aerobic bacteria.

oxides of nitrogen   [NO_x]   The sum of the concentrations of nitric oxide and nitrogen dioxide present in the ambient air as determined by employing an acceptable gas sampling procedure.

oxidizer   A gas or liquid that accelerates combustion, and that on contact with a combustible material may cause a fire or an explosion.

oxidizing gas   A gas that supports combustion, such as oxygen, a halogen, nitrous oxide, and others.

oxidizing material   Chemicals or chemical combinations that spontaneously evolve oxygen at room temperature or with slight heating. *See* oxidizer.

oxygen deficiency   An oxygen concentration in air of less than 20.8 percent. Some regulatory agencies typically define an oxygen deficiency as a level below that of oxygen in the normal atmosphere, while others identify it at some level below that (e.g., 19.5 percent).

oxygen deficiency—immediately dangerous to life or health   An atmosphere that causes the partial pressure of oxygen in the inspired air to be equal to or less than 100 millimeters of mercury in the upper portion of the lungs. The partial pressure of oxygen in the atmosphere is typically 158 millimeters of mercury.

oxygen-deficient atmosphere (OSHA)   That concentration of oxygen by volume below which atmosphere-supplying respiratory protection must be provided. An oxygen deficient atmosphere exists where the percentage of oxygen by volume is less than 19.5 percent.

oxygen-enriched atmosphere (Fire Protection)   Any atmosphere in which the concentration of oxygen exceeds 21 percent by volume, or the partial pressure exceeds 160 millimeters of mercury, or both.

oxygen-enriched atmosphere (OSHA)   An atmosphere containing more than 23.5 percent oxygen by volume.

oxygen toxicity   A toxic effect resulting from inhaling air containing a high partial pressure of oxygen. A disease that occurs among divers and others who work in pressurized atmospheres.

oxyhemoglobin   Hemoglobin that has absorbed oxygen (i.e., has been oxygenated).

oz   Ounce.

ozonation   The use of ozone (rather than chlorine) to treat water for the removal of pathogenic organisms.

ozone   An allotropic form of oxygen.

ozone layer   The region of the upper atmosphere where most ozone is concentrated. Also referred to as the ozonosphere.

P

p Pico, $1 E^{-12}$.

p Peta, $1 E^{15}$.

Pa Pascal.

packed tower (Ventilation) Contact beds of stone, ceramic balls, or wood through which gases and liquid pass concurrently, counter current, or in cross flow for the removal of a gas, vapor, or mist in exhaust air.

PACM Presumed asbestos-containing material.

paddle wheel fan Centrifugal fan with radial blades.

PAH Polynuclear aromatic hydrocarbon or polycyclic aromatic hydrocarbon.

pain Generally a localized unpleasant or agonizing sensation that is the result of stimulation of sensory nerve terminals located in almost every part of the body.

pair production A process by which radiation loses energy to matter. It involves the creation of a positron–electron pair from a photon of at least 1.02 mega electron volts.

pallor Paleness, or absence of skin color.

palpitations Rapid action of the heart noted by the individual. May be regular or irregular.

palsy Uncontrollable tremor of the body or one or more body parts.

PAMS Photochemical assessment monitoring station.

pancytopenia A deficiency of all cell elements of the blood.

pandemic An epidemic over a wide geographical area, crossing international borders and usually affecting a large number of people.

P and IDs Piping and instrumentation drawings.

P and Ps Practices and procedures.

panic (Life Safety) Sudden, overpowering terror, often affecting many people simultaneously.

panic hardware Hardware designed to facilitate safe egress of people in case of an emergency when a pressure of not more than 15 pounds is applied to the releasing device in the direction of exit travel.

papilloma A small growth or benign tumor of the skin or mucous membrane.

PAPR Powered air-purifying respirator.

papular Characterized by a papule, which is a small, superficial, solid elevation of the skin.

papule Small raised red, brown, yellow, white, or skin-colored skin lesion.

paracusis The hearing of sound at a different frequency than that at which it exists.

paradigm A typical example or pattern of thought regarding phenomena with which scientists usually work.

paralysis Temporary or permanent impairment, or the loss of voluntary motor function, that is generally accompanied by a loss of feeling in part of the body.

paralytic   Individual who is suffering from paralysis.

paramagnetic   Attracted into a magnetic field. Oxygen is paramagnetic at normal temperatures; this property is a basis for measuring oxygen concentration.

parameter   A characteristic of a population, such as the mean, standard deviation, or the variance. Also a variable quantity or arbitrary constant appearing in a mathematical expression, each value of which restricts or determines the form of the expression.

paranoia   Chronic mental disorder in which the individual suffers from delusions.

paraoccupational exposure   An exposure that occurs when workers are exposed to contaminants in the workplace and carry them outside the worksite on their clothing, body, or by other means. As a result, nonworkers such as family members can receive an exposure to the contaminant.

paraparesis   Mild paralysis of the lower extremities.

paraplegia   Paralysis of the legs and lower part of the body.

parasite   A plant, animal, or microbiological organism that lives upon or within another living organism at whose expense it obtains some advantage. It does not necessarily cause disease.

parasthenia   A condition of organic tissue causing it to function at abnormal intervals.

parenchyma   The essential or functional elements of an organ.

parent   A radionuclide that, upon disintegration, yields a specified nuclide either directly or as a later member of the radioactive series.

parent compound   A chemical or metabolically unchanged compound.

parenteral   Substance introduced into the body by a route other than by way of the intestines, such as through the skin.

paresthesia   A morbid, abnormal, or perverted sensation.

Parkinson's Law   A law that states that work will expand to use all available resources.

paroxysm   The sudden recurrence or increase in the severity of a disease.

partial body irradiation (Ionizing Radiation)   The exposure of only a part of the body to incident radiation energy.

partial disability   A condition in which a person can work, but has some limitations and cannot do his/her customary work.

partial pressure   That part of the total pressure of a mixture of gases that is contributed by one of the constituents. In any gas mixture the total pressure is equal to the sum of the pressures each gas would exert if it were alone in the volume occupied by the mixture.

particle   A small discrete mass of solid or liquid matter.

particle size   The measured dimension of liquid or solid particles, usually expressed as the diameter in microns.

particle size distribution   The statistical distribution of the size or mass of an aerosol, typically described by the geometric mean and standard deviation of the distribution. The data are useful in estimating aerosol exposures to various regions of the respiratory tract.

Particle Size-Selective TLVs   Exposure limits that recognize the size-fraction most closely associated for each substance with the health effect of concern, and the mass concentration within that size-fraction that should represent the TLV.

particulate   A liquid or solid particle. When suspended in air, they are referred to as aerosols; when on a surface, they are referred to as settled particulates. Settled particulates, such as dust and fumes, can be resuspended in the air.

particulate matter   [PM]   A suspension of solid or liquid particles in air. Also referred to as an aerosol.

particulates not otherwise classified   [PNOC]   Particulates for which there is no evidence of specific toxic effect or that do not cause fibrosis or systemic effects, but are not biologically inert. This recognizes that all materials are potentially toxic.

partition   The tendency of a substance to exhibit an affinity for one material over another.

parts per billion   [ppb]   Unit of concentration in which one part of a contaminant is present in one billion parts of air or other media by volume.

parts per million   [ppm]   An expression of concentration as the number of parts of a contaminant in one million parts of air or other media.

parts per trillion   [ppt]   An expression of concentration in which one part of a contaminant is present in one trillion parts of air or other media.

pascal   [Pa]   The metric unit of pressure measurement equal to a force of 1 newton acting on an area of 1 square meter.

passive   Inactive.

passive immunity   Acquired immunity in a person that results from the administration of preformed antibodies or specifically sensitized cells.

passive sampling   A sampling methodology in which air containing a contaminant penetrates through a semipermeable membrane, where it is either adsorbed on a solid sorbent, absorbed in a liquid sorbent, or detected by a passive-type detector (e.g., electrochemical, catalytic, etc.). Also called diffusive sampling.

passive smoking   The inhalation of tobacco smoke by nonsmokers as a result of the smoke exhaled by smokers contaminating the air breathed by nonsmokers.

pasteurization   A process of heating a liquid to a specified temperature for a definite time period to kill most species of pathogenic organisms and to delay bacterial development.

patch testing   The taping of suspected allergens to the skin and observing the appearance of irritation at specific time intervals, typically at 48 hours. Tests must be performed with controls to ensure the skin is not responding to the tape or the skin covering.

pathogen   A disease-producing microorganism or material.

pathogenesis   The cellular events, reactions, and other pathologic mechanisms that occur in the development of disease.

pathogenic   An agent, usually infectious, that is capable of causing disease.

pathogenic bacteria   Bacteria that may cause disease or morbid symptoms in the host organism by their parasitic growth.

pathogenicity   Ability of an agent to cause pathologic changes or disease.

pathogenic waste   Waste that may contain organisms capable of causing disease.

pathological   Abnormal or diseased.

pathologist   A licensed physician who specializes in the study of the origin, nature, and course of diseases.

pathology   A branch of medicine that determines the essential nature of diseases, especially of the structural and functional changes in tissue and organs of the body that cause or are caused by disease.

pathway   The course followed in the attainment of a specific end, such as how a pathogenic organism gets to man.

patient   Individual who is ill, has a disease, is suffering, or is under treatment by a health care professional.

PAT sample   Proficiency Analytical Testing sample from NIOSH used to assess performance of those laboratories involved in the analysis of workplace air samples and who want to be accredited, or to maintain accreditation, by the American Industrial Hygiene Association.

PBB   Polybrominated biphenyl.

PBZ   Personal breathing zone.

PCB   Polychlorinated biphenyl.

PCB article   Any manufactured article that contains PCBs and whose surfaces have been in direct contact with PCBs (excludes PCB containers).

PCB-contaminated transformer   A transformer that contains 50 parts per million or greater of PCBs, but less than 500 parts per million.

PCB item   Any PCB article, PCB article container, or PCB equipment that contains PCBs at a concentration of 50 parts per million or greater.

PCB transformer   A transformer that contains 500 parts per million PCBs or greater.

pcf   Pound(s) per cubic foot.

pCi   Picocurie(s).

pCi/L   Picocuries per liter.

PCIH   Professional Conference on Industrial Hygiene.

PCM   Phase contrast microscopy.

pct   Percent.

peak concentration   The concentration of an airborne contaminant that may be much higher than average and typically occurs for only short periods of time.

peak exposure   The highest concentration that occurs/occurred during a sampling period.

peak noise level   The maximum instantaneous sound pressure level that occurs for a short duration or in a specified time interval.

peer review   The evaluation by peer practitioners and other professionals of the effectiveness, quality, and relevance of professional studies that have been completed, documented, and submitted for publication. In industrial hygiene, it is the process of the review of manuscripts submitted for publication which are judged for scientific and technical merit by other industrial hygienists.

PEL (OSHA)   Permissible exposure limit.

PEL–C (OSHA)   Permissible exposure limit–ceiling.

PEL–STEL (OSHA)   Permissible exposure limit–short term exposure limit (typically 15-minute exposure).

PEL–TWA (OSHA)   Permissible exposure limit–time-weighted average (i.e., 8-hour).

penetrating encapsulant   A liquid material that is applied to asbestos-containing material to control the release of asbestos fibers by bonding the material together.

penetration (Protective Clothing)   The passage of a chemical through seams, pinholes, closures, or other imperfections in a material used for protective clothing.

peptic ulcer   Stomach ulcer.

perceived noise level   The noise level in decibels assigned to a noise by means of a calculation procedure that is based on an approximation to subjective evaluations of noisiness.

percent bias   *See* bias.

percentile   A score in a distribution, below which falls the percent of scores indicated by the stated percentile. For example, a score of 91 percent would be in the 90th percentile.

perception   The conscious mental awareness of a sensory stimulus.

percolation   The downward flow or filtering of water through spaces in soil or rock.

percutaneous   Administered or absorbed through the unbroken skin, such as the absorption of a hazardous material (e.g., phenol, hydrazine, etc.).

percutaneous toxicity   The ability of a substance to be absorbed through the skin and cause a toxic effect.

performance-oriented standard   A standard that outlines the level of performance that must be demonstrated. It provides flexibility to the employer to develop a compliance strategy that is reflective of the needs of the facility; that is, it allows the employer to choose the most appropriate pathways to achieve compliance.

pericardium   The sac that surrounds the heart.

perinatal   Pertaining to or occurring in the period shortly before and after birth.

period (Vibration)   The time required for a complete oscillation or for a single cycle of events.

periodicity   The repetitive characteristics of a signal.

periodic table   A systematic classification and arrangement of the chemical elements according to their atomic numbers and their physical and chemical properties.

periodontitis   Inflammation of the gums.

peripheral   Situated away from a center or central structure.

peripheral nervous system   The nerves outside the central nervous system that control muscles and glands and that receive sensory signals.

peripheral neuritis   Inflammation of peripheral nerves.

peripheral neuropathy   *See* peripheral polyneuropathy.

peripheral polyneuropathy   A progressive and potentially irreversible disorder of the peripheral nervous system. Also referred to as peripheral neuropathy. N-hexane is one substance that has caused this disease.

peristalsis   The method by which the digestive tract propels its contents forward.

peritoneum   The membrane that lines the abdominal cavity and the pelvic region.

peritonitis   Inflammation of the peritoneum, resulting in abdominal pain, tenderness, constipation, vomiting, and fever.

permanent disability   Permanent impairment, including any degree of impairment, such as amputation of a finger, impairment of vision, or other permanent, crippling, nonfatal injury.

permanent partial disability   A condition in which there is presently a partial loss of earning power, and the injured person's future earning capacity has also been affected.

permanent total disability   A condition in which the worker's wage-earning capacity has been affected such that the injured cannot compete in the job market.

permanent threshold shift   [PTS]   A permanent lessening of an individual's ability to hear.

permeability   The passage of a fluid or dissolved solid through a membrane, cell, or other semipermeable membrane.

permeation   Process by which a chemical substance moves through a material, such as gloves, clothing, etc.

permeation rate   The rate at which a chemical substance moves through a material, such as gloves, clothing, etc.

permissible dose (Ionizing Radiation)   The dose of ionizing radiation that may be received by an individual within a specified period without harmful effect.

permissible exposure limit   [PEL]   An exposure limit that is set for exposure to a hazardous substance or harmful agent and enforced by OSHA as a legal standard. Typically based on time-weighted average concentrations for a normal 8-hour workday and 40-hour workweek.

permissible exposure limit–ceiling (OSHA)   [PEL–C]   Concentration of a substance to which a worker may be exposed and which shall not be exceeded.

permissible exposure limit–short-term exposure limit (OSHA)   [PEL–STEL]   A 15-minute, time-weighted average exposure limit that is not to be exceeded during the workday.

permissible exposure limit–time-weighted average (OSHA)   An 8-hour, time-weighted average exposure limit that must not be exceeded in the 8-hour work shift.

permit   A document authorizing specific work to be carried out at a specific location if specific procedures/practices are adhered to. With respect to EPA regulations, it provides information as to the limitations, requirements, and restrictions applicable to discharges, emissions, and disposal practices; compliance schedules; and monitoring, analytical, and reporting requirements.

permutation   Any of the possible combinations or changes in position within a group.

pernicious   Tending to cause death or serious injury.

pernicious anemia   A severe form of blood disease marked by the progressive decrease in red blood cells, muscular weakness, and gastrointestinal and neural disturbances.

per os   Orally, or by mouth.

peroxyacylnitrates   [PAN]   Organic compounds that are formed in the atmosphere from other pollutants. They cause eye irritation and respiratory distress.

persistent   Durable, or refusing to give up or let go.

persistent environmental chemical   A chemical that decomposes slowly, if at all, in the environment.

persistent pesticide   Pesticide that does not break down in the environment and that remains for some time after its application.

personal breathing zone   [PBZ]   The region of the air from which a worker inhales in the normal carrying-out of a work assignment.

personal eyewash   A supplementary eyewash that supports plumbed units, self-contained units, or both, by delivering immediate flushing of the eyes for less than 15 minutes.

personal monitoring   A type of environmental monitoring in which an individual's exposure to a substance or agent is determined. *See* also personnel monitoring and personal sample.

personal noise dosimeter   An instrument designed to be worn by a worker for determining the individual's noise exposure while performing a specific task or over the entire 8-hour workday.

personal protective equipment   [PPE]   Equipment worn by workers to protect them from exposure to hazardous materials or physical agents that they may encounter in the work environment.

personal sample   A sample taken in the breathing zone or other area of a person (i.e., at the ear, body, etc.) to determine the potential for an adverse health effect to occur to the individual as the result of exposure to an airborne contaminant, physical agent, etc.

personnel monitoring   Monitoring any part of an individual; that is, the breathing zone, the area where the individual works, his/her excretions, clothing, etc. *See* personal sample.

personnel monitoring device   Device designed to be worn or carried by an individual for the purpose of measuring the dose of radiation received or the amount of physical agent or airborne contaminant to which a person is exposed.

person–sievert   A means of expressing the collective dose of ionizing radiation to a population; the product of the average dose times the number of people exposed.

person–years (Epidemiology)   The product of the number of persons under observation and the duration of observations of each person.

perspiration   The moisture excreted through the pores of the skin by the sweat glands.

perturbation   The state or condition of being agitated; or a disturbance to the state of equilibrium.

pesticide   Any chemical that is used to kill pests, especially insects and rodents.

pesticide types   Botanicals (e.g., pyrethrins), carbamates (e.g., aldicarb, carbaryl, et al.), organohalides (e.g., aldrin, DDT, lindane, etc.), and organophosphates (melathion, parathion, etc.).

peta   [P]   Prefix denoting 1 E^{15}.

petri dish   A transparent dish and cover used in the culturing of microorganisms and tissues.

petrochemicals   Chemicals derived from petroleum or natural gas, including ammonia and thousands of organic chemicals.

PF   Protection factor.

PFD   Personal flotation device.

PFD   Process flow diagram.

pg   Picogram, 1 E $^{-12}$ g.

pH   Term expressing the acidity or alkalinity of a solution, with neutral indicated at a pH of 7. The logarithm of the reciprocal of the hydrogen-ion concentration.

PHA   Process hazard analysis.

phagocyte   Any cell in the body that engulfs/ingests microorganisms or other cells and foreign material. Fixed phagocytes are potentially phagocytic, and free phagocytes are intensely phagocytic.

phagocytosis   The envelopment and digestion of bacteria and other foreign bodies by phagocytes.

phalanges   Any of the bones of the fingers or toes.

pharmacokinetics   The study of the action of a substance on the body over a period of time, including the processes of its absorption, distribution, localization in tissues, biotransformation, and excretion.

pharynx   Th muscular membrane sac between the mouth and nares and the espohagus.

phase contrast illumination   An optical arrangement in a microscope in which one part of the illumination is treated differently from the rest, allowing the imaging of transparent objects such as fibers.

phase contrast microscope   An optical microscopic technique for determining the concentration of fibers in an air sample. The method cannot distinguish the types of fibers that are present. This technique enables the microscope to transfer differences in the phase of light waves into intensity variations that increase specimen contrast, thereby enabling a specimen to be seen that would otherwise be essentially invisible by light field microscopic techniques.

pheromone   A substance secreted to the outside of the body by an individual that is perceived by and elicits a behavioral response from a second individual of the same species.

phlebitis   Inflammation of a vein.

phlebotomist   One trained to remove blood from blood vessels.

phlebotomy   The removal of blood from blood vessels.

phlegm   Stringy, thick mucus secreted by the respiratory mucosa.

phobia   Persistent, abnormal fear of a specific thing or situation; often an illogical fear.

phon   Unit of loudness level; a unit of subjective loudness that is based on a decibel scale.

phosphor   A liquid or crystalline, organic or inorganic substance that is capable of absorbing energy (e.g., X-rays, ultraviolet radiation, etc.) and emitting a portion of the energy in the visible, infrared, or ultraviolet region of the electromagnetic spectrum.

phossy jaw   An occupational disease resulting from exposure to white phosphorous.

photoallergy   An allergic reaction that is a heightened, delayed, contact-type sensitivity to light. An allergic reaction caused by exposure to a chemical that is absorbed into the body and subsequently activated by ultraviolet radiation, with resulting effects of irritation or allergic contact dermatitis.

photochemical oxidants   Air pollutants formed by the action of sunlight on oxides of nitrogen and hydrocarbons.

photochemical reaction   The chemical changes that are induced as a result of the absorption of radiant energy (e.g., light) by various substances.

photochemical smog   Air pollution condition that is a result of atmospheric chemical reactions of various pollutants including, principally, the oxides of nitrogen and hydrocarbons.

photodermatitis   A skin problem due to a chemical that is activated by light. A number of chemical substances do not directly cause irritation on contact with the skin, but symptoms appear when the skin is exposed to sunlight.

photodosimetry (Ionizing Radiation)   A determination, through the use of photographic film referred to as a film badge, of the cumulative dose of ionizing radiation received by a person.

photoelectric effect   A process by which radiation loses energy to matter. All the energy of a photon is absorbed in ejecting an electron from the material/substance and imparting kinetic energy to the electron.

photoionization detector   Photons of light energy from an ultraviolet lamp in an instrument are absorbed by some molecules/species and dissociation results, producing ions and electrons. The amount of dissociation that occurs is proportional to the contaminant concentration in the sampled air.

photokeratitis   Inflammation of the cornea as a result of exposure to ultraviolet light. A feeling of sand in the eyes. Often experienced by welders.

photolysis   The decomposition of a compound as a result of the absorption of radiant energy. Also referred to as photochemical decomposition.

photometry   Analytical method based on the determination of the relative radiant power of a beam of radiant energy in the visible, infrared, or ultraviolet region of the electromagnetic spectrum that has been attenuated as a result of its passing through a solution or gas–air mixture containing a material that can absorb the radiant energy. Also referred to as colorimetry.

photomultiplier tube   A vacuum tube capable of increasing (multiplying) the electron input to the tube.

photon   A quantity of electromagnetic energy having the characteristics of a particle.

photophobia   Abnormal visual intolerance to light.

photosensitivity   The sensitivity of an organism or chemical to light.

photosensitization   Dermatitis due to exposure to a sensitizer followed by exposure to light, with resulting photocontact dermatitis. The development of abnormally heightened reactivity of the skin to sunlight.

photosensitizer   A substance that increases the sensitivity to light.

photosynthesis   The utilization of sunlight by plants as well as bacteria to convert two inorganic substances (carbon dioxide and water) into carbohydrates. This is an example of a photochemical reaction.

phototoxic   The capacity of a substance to sensitize the skin to a harmful light-induced reaction; erythema followed by hyperpigmentation of unexposed areas of the skin resulting from exposure to agents containing photosensitizing substances, such as coal tar and some drugs, then to sunlight.

phototoxin   A phototoxic substance.

physical change   A change in the state of matter, but not such that a new substance is produced.

physical chemical hazard   A chemical for which there is scientifically validated evidence that it is a combustible liquid, a compressed gas, explosive, flammable, an organic peroxide, an oxidizer, pyphoric, unstable, or water-reactive.

physiological half-life   The time required for the body to eliminate one-half the amount of a material absorbed or ingested.

physiology   The science that addresses the functions of the living organism and its parts, the physical and chemical factors and processes involved, and the study of the qualitative and quantitative aspects of those processes necessary to the function of the living body.

phytodermatitis   Dermatitis resulting from exposure to plants.

phytoremediation   The use of plants and trees to remove, degrade, or sequester hazardous materials from contaminated soil or groundwater.

phytotoxic   Something that harms plants.

pica   A craving to ingest nonfood items, or an abnormal appetite, such as that of children consuming paint chips containing lead.

pico   [p]   One-trillionth, $1 E^{-12}$.

picocurie   [pCi]   A unit of measurement of radioactivity equal to one-trillionth of a curie, $1 E^{-12}$ Ci.

picogram   [pg]   One-trillionth of a gram, $1 E^{-12}$ g.

PID   Photoionization detector.

piezoelectric   A material that provides a conversion between mechanical and electrical energy.

pig (Ionizing Radiation)   A container, typically constructed of lead, that is used to ship or store a radioactive material.

pig (Petroleum Industry)   A jointed metal device which can be forced through a pipe line by hydraulic pressure to scrape off rust and scale or to mark the interface between two products being transferred through the pipe line.

pinch bar   A steel lever having a pointed projection at one end for use in moving heavy loads.

pinch point (Machine)   Any point, other than the point of operation, at which it is possible for a part of the body to be caught between the moving and stationary parts of equipment.

pink noise   Noise whose noise-power per unit frequency is inversely proportional to frequency over a specified range. Noise that decreases with increasing frequency to yield constant energy per octave band.

pinna   The outer funnel-like structure of the ear.

pint   A unit of volume equal to 473 milliliters.

pipe lagging   The insulation or wrapping around pipe.

pipette   A basic piece of laboratory equipment that is used for volumetric measurement of fluids and for the transfer of those fluids from one container to another.

piscicide   A substance that is poisonous to fish.

pitch (Acoustics)   The attribute of auditory sensation in terms of which sounds may be ordered on a scale extending from low to high. Pitch depends primarily on the frequency of the sound stimulus, but also on the sound pressure and wave form of the sound.

pitot traverse   A series of measurements employing a pitot tube at predetermined positions across a section of ductwork or piping in order to determine total, static, and velocity pressures for subsequent use in determining air velocity in the duct/pipe and the amount of air passing the point at which the pitot traverse was made.

pitot tube   A device for measuring pressures in a ventilation system consisting of two concentric tubes arranged so that one measures total or impact pressure, and the other measures static pressure. The difference in pressure indicated on a U-tube connected between the total pressure and static pressure tubes represents the velocity pressure at that position in the duct.

pixels   Dots that form the picture on a CRT screen.

placard   Information presented on a sign for display so that it is visible to the public. Typically used on trucks and rail cars to alert individuals of the material being transported.

placebo   An inactive substance or preparation given to satisfy an individual's need for therapy, as well as for use in controlled studies to determine the efficacy or effect of a substance or medicinal product.

plankton   Small plants and animals that live in water.

plaque   A patch or flat area on a surface, such as the skin or lining of a blood vessel.

plasma   The fluid part of the blood in which the blood cells are suspended.

platelet   Disk-shaped structures found in the blood of all mammals and chiefly known for their role in blood coagulation.

plenum   A low-velocity chamber in a ventilation system that is employed to distribute static pressure throughout its interior. An air compartment or chamber to which one or more ducts are connected and that forms part of an air distribution system.

plenum velocity   The air velocity within a plenum.

pleura   A thin membrane surrounding the lungs and lining the internal surfaces of the chest cavity. The pleura reduces the friction of the movements of the lungs, chest, etc. during respiration.

pleural plaques   Plaques observed in the pleura of some of the persons who have been exposed to asbestos.

pleurisy   Irritation, inflammation, and pain of the outer lung lining and the chest cavity's inner lining.

Plimsoll mark   A marking placed on the side of a ship denoting the maximum depth to which it may be loaded or ballasted.

PLM   Polarized light microscopy.

plumbism   Lead poisoning.

plume   The visible emissions from a single point of origin, such as a flue, stack, or chimney.

plutonism   Disease resulting from plutonium exposure.

PM   Particulate matter.

PM   Preventive maintenance.

PM 10   Particulate matter with an aerodynamic diameter less than or equal to 10 micrometers.

PMN   Premanufacture notification.

PNA(s)   Polynuclear aromatic compound(s).

pneumatic   Operated by air pressure.

pneumoconiosis   A condition, usually of occupational or environmental origin, characterized by the permanent deposition of substantial amounts of particulate matter in the lungs and by tissue reaction to its presence. It may be a relatively harmless form, such as siderosis, or a serious form, such as silicosis.

pneumoconiosis-producing dust   A dust that, when inhaled, deposited, and retained in the lungs, produces signs, symptoms, and findings of pulmonary disease.

pneumonitis   Inflammation of the lungs.

pneumonoultramicroscopicsilicovolcanokoniosis   A special form of silicosis caused by ultra-microscopic particles of siliceous volcanic dust. This is the longest word I have ever seen in the IH/safety literature (from "Dust Producing Operations in the Production of Petroleum Products," July 1937, in the files of Standard Oil Co., NJ).

PNOC   Particulates not otherwise classified.

PNS   Peripheral nervous system.

p.o.   Per os, or orally.

pocket dosimeter   A device for determining the dose of ionizing radiation received by a person. Also referred to as a pocket chamber.

point of operation (Machinery)   The point at which cutting, shaping, or forming is accomplished upon the stock, including other points that may offer a hazard to the operator.

point source   A stationary location, usually industry, from which pollutants are released into the environment.

point-source detector   Single point detection device that responds to a contaminant as it is transported by air currents from its source to the detector location.

poise   The unit of viscosity of a liquid, defined as the force in dynes required to move a surface 1 square centimeter in area past a parallel surface at a speed of 1 centimeter per second, with the surfaces separated by a fluid film 1 centimeter thick. The commonly used unit is the centipoise, which is 1/100 of a poise.

poison (Catalyst Technology)   Any compound that causes a catalyst to lose activity.

poison (Toxicology)   Any substance that, when ingested, inhaled, or absorbed, or when applied to, injected into, or developed within the body, in relatively small amounts, may cause damage to the structure of or disturbance to the function of the body by its chemical action.

poisoning   The morbid or unhealthy effect produced by a poison.

poisonous   A substance that has poisonous properties and acts as a poison.

polar compound   A molecule in which the positive and negative electrical charges are permanently separated, as opposed to nonpolar molecules in which the charges coincide. Polar molecules ionize in solution and impart electrical conductivity. Examples of polar compounds are alcohol, water, sulfuric acid, etc.

polarized light   Light waves whose vibrations occur in one direction only.

polarized light microscopy   [PLM]   An optical microscopic technique used to distinguish different types of fibrous materials by their unique optical properties when exposed to polarized light.

polarography   An analytical method, based on the electrolysis of a sample solution, for determining the amount of specific contaminants that are electro-reducible or electro-oxidizable.

polar solvent   Solvent that contains oxygen.

policy statement (Safety/Industrial Hygiene)   A document with a statement of intent that identifies realistic, attainable goals, those responsible for specific aspects of a program/objective, the method of accountability for management and employees, the means to measure performance, and that has the full support of management.

pollen   The fine, powderlike material produced by plants that serves as the male element in the fertilization of plants.

pollutant (Environmental)   A contaminant released into the environment.

pollution   Contamination of the soil, water, or atmosphere beyond normal or natural levels, producing undesired environmental effects. The presence of matter whose nature, location, or quantity produces undesired environmental effects.

polyaromatic hydrocarbon   [PAH]   A group of organic compounds with ring structure that are highly reactive, some of which are carcinogenic.

polychlorinated biphenyls   [PCBs]   Chemical compounds used in electrical equipment as a coolant and, in some equipment, as a heat transfer media or hydraulic fluid.

polycyclic (Organic Chemistry)   Compound that has two or more rings in its molecular structure.

polycythemia   A condition marked by an excess of red blood cells in the blood.

polydisperse aerosol   An aerosol with a geometric standard deviation greater than 1. As the geometric standard deviation increases, the aerosol becomes more polydisperse.

polymer   *See* polymerization.

polymer fume fever   An occupational disease characterized by chills, dry cough, and tightness of the chest, that results from exposure to the breakdown products (due to heating) of fluorocarbons, such as polytetrafluoroethylene.

polymerization   A process in which the molecules of one compound link together into a larger unit referred to as a polymer containing from two (i.e., dimer) to hundreds of molecules.

polyneuropathy   A neuropathy affecting a number of peripheral nerves.

polynomial   Consisting of more than two names or terms.

polynuclear aromatic compounds   *See* polynuclear aromatic hydrocarbons and polyaromatic hydrocarbons.

polynuclear aromatic hydrocarbon   [PAHs]   Aromatic compounds containing three or more closed rings, usually of the benzene type. Also referred to as PNAs.

polyuria   Frequent discharge of urine, or a sharp increase in urination.

Pontiac fever   A short, febrile illness without pneumonia, characterized by headache, chills, cough, tiredness, muscle pain, and nausea, caused by Legionella bacteria and referred to as nonpneumonic Legionnaire disease.

population   The total group of individual persons, objects, or items from which samples may be taken to estimate characteristics of that population by statistical methods.

population parameters (Epidemiology)   The true parameters that are determined by including the entire population in an epidemiology study.

population risk   The excess number of cases of a disease in an exposed population above the expected number.

porosity   The property of being porous or admitting the passage of a gas or liquid through pores or interstices.

porphyrin   Any of a group of iron-free or magnesium-free cyclic tetrapyrrole derivatives that occur universally in protoplasm. Protoporphyrin (zinc protoporphyrin is a test in determining inorganic lead absorption) is among them.

portable (Instrument)   A self-contained, battery-operated instrument that weighs less than 10 pounds and can be carried and used by an individual.

portable direct-reading instrument   A direct-reading, portable instrument that can directly measure the concentration of gases, vapors, or other contaminant, or the level of physical stress (i.e., noise, ionizing radiation, etc.).

portal of entry   Avenue (e.g., via inhalation, skin absorption, etc.) by which an agent (e.g., parasite, chemical, etc.) enters the body.

Porton reticle   A transparent grid that is mounted in the eyepiece of a microscope at the exact focal plane of the specimen and thus superimposed on the field being viewed, thereby facilitating the sizing of particles collected on a filter or other collecting media.

position sensitivity (Instrument)   The effect on an instrument's response due to deviations in attitude from the normal operating position.

positive pressure   Condition that exists when more air is supplied to a space than is exhausted, causing the air pressure within that space to be greater than that in the surroundings. This condition can exist in a duct, room, building, etc.

positive-pressure respirator   A respiratory protective device in which the air pressure inside the respirator air inlet is positive in relation to the air pressure of the outside atmosphere during exhalation and inhalation.

positron   A particle that has the same weight as an electron, but that is electrically positive rather than negative.

posterior   Located behind a part or toward the rear of a structure.

post-traumatic stress disorder   [PTSD]   A specific anxiety condition that occurs following a stressful or traumatic event.

posture   The position or attitude of the body or bodily parts, such as the characteristic way one stands, sits, reclines, carries the head, etc.

potable   Safe to drink.

potable water (OSHA)   Water that meets the quality standards prescribed in the U.S. Public Health Service Drinking Water Standards, or water that is approved for drinking water purposes by the state or local authority having jurisdiction.

potency (Toxicology)   The ability of a contaminant or physical agent to produce an adverse health effect. The amount of a chemical that will produce an effect.

potent   Powerful or strong.

potential energy   Energy due to the position of an object rather than to its motion.

potential hazard   A situation that possesses characteristics conducive to the occurrence of an exposure to a hazardous agent, physical stress, ergonomic stressor, or other hazard.

potentiation   The enhancement of the action of a substance by another, such that the effect produced is greater then the sum of the effects of each alone. *See* synergistic.

potentiator   A chemical that has little adverse effect itself; however, when given or received in conjunction with another chemical, it enhances the effect of that chemical.

potentiometer   Instrument for measuring an unknown voltage. The potential difference is determined by comparison to a standard voltage.

potter's asthma   Asthmatic symptoms associated with the pneumoconiosis observed among workers in the ceramic industry.

POTW   Publicly owned treatment works.

pound   A unit of weight equal to 453.5 grams.

pound mole   The amount of a substance, in pounds, that is equivalent to the molecular weight of the substance. For example, a pound mole of sodium hydroxide is equal to 40 pounds.

pounds per square inch absolute   [psia]   A unit of pressure measurement with zero pressure as the base or reference pressure.

pounds per square inch gauge   [psig]   A unit of pressure measurement with atmospheric pressure as reference pressure, such that zero gauge pressure is equal to 14.7 psia.

pour point   The lowest temperature at which a liquid will flow.

power   The time rate at which work is done.

power (Epidemiology)   The ability of a study to demonstrate an association if one exists.

power density   The rate of energy transported into a small sphere divided by the cross-sectional area of that sphere, expressed in units of watts per meter squared (W/m^2), or more commonly, as milliwatts per square centimeter (mW/cm^2).

powered-air purifying respirator   [PAPR]   A respiratory protective device that has air under pressure provided to the wearer by a fan/pump after it has been cleaned by drawing it through a filter or chemical cartridge.

power level (acoustics)   Ten times the logarithm to the base 10 of the ratio of a given power to a reference power. The reference power is typically taken as 1 E^{-12} watts.

PPA   Personal protective apparel.

ppb   Parts per billion.

ppbv   Parts per billion by volume.

ppcf   Particles per cubic foot.

PPD   Purified protein derivative. For example, a PPD of tuberculin is used in screening people for tuberculosis.

PPE   Personal protective equipment.

ppm   Parts per million.

ppm-hr   Parts per million-hours.

ppmv   Parts per million by volume.

ppmw   Parts per million by weight.

ppt   Parts per trillion.

pptv   Parts per trillion by volume.

PRCS   Permit-required confined space.

precautions   Measures taken to reduce the likelihood of an excessive exposure to a health hazard.

precipitate   A solid that is separated from a liquid as the result of a chemical or physical change.

precipitation   All forms of water particles, liquid or solid, that fall from the atmosphere and reach the earth.

precision   The agreement among repeated measurements of the same parameter under the same conditions.

preclude   To make impossible by some previous action.

precursor   Something that precedes or exists before another is formed. A condition or state preceding pathological onset of disease.

predisposing factors   Factors such as age, sex, weight, skin color, health status, etc. that may increase an individual's susceptibility to a potential hazard.

preemployment screening (Hearing)   The determination of an employee's hearing ability prior to his/her entering an environment where there is exposure to noise.

preexisting disease   A disease known to exist before the onset of current symptoms.

premanufacture notification   [PMN]   A notice that must be made to the EPA when a company intends to manufacture or import a new chemical or when a company intends to develop a significant new use for a chemical substance.

premise   A proposition upon which an argument is based.

premises   The land and buildings upon it.

prenatal   Preceding birth.

presbycusis   The condition of hearing loss specifically ascribed to aging; hearing loss due to aging.

presbyopia   Nearsightedness that typically accompanies advancing age.

prescription   A written directive to compound, dispense, or administer medication or services to an individual.

preservative   Substance added to a product to prevent spoilage or inhibit the growth or multiplication of microorganisms.

pressure   The normal force exerted by a fluid or gas per unit area on the walls of its containment, expressed as the force per unit area, such as pounds per square inch.

pressure-demand respirator   A respiratory protective device in which the air pressure inside the facepiece is greater than atmospheric at all times, and that admits additional breathing air when the pressure in the facepiece is reduced by inhalation.

pressure drop   The difference in the static pressure when measured at two positions in a ventilation system. The difference is due to friction and turbulence to which the air is subjected in its transport through the ventilation system. Typically measured in inches of water pressure.

pressure loss   The energy loss associated with the movement of air through a ventilation system as a result of friction and turbulence. Typically measured in inches of water. *See* pressure drop.

pressure relief valve   A safety valve held closed by a spring or other means and designed to automatically relieve pressure in excess of its pressure setting.

pressure vessel   A leak-tight pressure container, usually cylindrical or spheroidal in shape, designed to contain liquids, with pressure substantially greater than atmospheric (e.g., from 15 psi to 5000 psi).

presumed asbestos-containing material   [PACM]   Material that is assumed to be asbestos-containing without testing it to determine the presence of asbestos at 1 percent or more by weight.

prevailing wind   The predominant direction from which the wind blows.

prevalence   The number of cases of disease, infected persons, or persons with some other attribute that are present at a particular time and in relation to the population from which they are drawn.

prevalence rate   The ratio of the number of cases of some condition at one point in time to the total population at risk at that time. The prevalence rate is often expressed as a percentage.

prevention (Public Health)   The act of promoting, preserving, and restoring health when it is impaired; minimizing suffering and distress.

prevention of significant deterioration   [PSD]   A policy of the CAA that limits increases of air contamination in clean air areas to certain increments even though air quality standards are being met.

preventive maintenance   [PM]   Scheduled overhaul or repair.

prickly heat   A condition due to obstruction of the ducts of the sweat glands, probably as the result of irritation of the skin surface. Characterized by skin reddening, itching, and swelling.

prima facie   Presumed to be true before closer inspection.

primary air pollutant   Pollutants released directly to the atmosphere, typically as the result of an industrial or combustion process.

primary calibration (Instrumentation)   A calibration procedure in which the instrument output is observed and recorded while the input stimulus (sample) is applied under precise conditions, usually from a primary standard traceable to the National Institute of Standards and Technology.

primary calibration method (Flow Rate)   A calibration method for determining sample pump flow rate that is generally a direct measurement of volume on the basis of the physical dimensions of an enclosed space. The use of a spirometer, Mariotti bottle, or a soap bubble meter are primary calibration methods for pumps.

primary carcinogen   A direct-acting carcinogen that does not need metabolic activation.

primary combustion air   Air that is first introduced to a combustion system, such as a furnace burner, when the fuel is first burned.

primary drinking water standard   A standard that identifies and specifies the levels of contamination that are acceptable for public drinking water, the purpose of which is to prevent adverse health effects.

primary irritant   A substance that produces a recognized irritating effect at the location of skin contact. Primary irritants affect everyone, but all primary irritants do not produce the same degree of irritation.

primary pollutant   A pollutant emitted directly from a polluting source.

primary skin irritant   A material that acts directly on the skin, disturbing membrane structure and affecting the osmotic pressure of skin cells.

primary standard (Air Pollution)   A national (U.S.) primary ambient air quality standard promulgated under the Clean Air Act. It is a level of air quality that will protect public health.

primary standard (Flow Rate)   A device that enables the direct measurement of the volume of air flow on the basis of the physical dimensions of an enclosed space, such as by use of a spirometer, Mariotti bottle, or soap bubble meter. Such devices have no working parts and are not subject to corrosion or friction to any extent.

primary treatment   A first stage in water treatment in which floating or settleable solids are removed.

primate   An individual belonging to the order Primates, which includes man, apes, monkeys, and lemurs.

probability   The likelihood that an event will occur, expressed quantitatively; an event that can reasonably be expected to occur on the basis of available evidence.

probable cause (Law)   Reasonable grounds for a belief that an accused person is guilty as charged.

probe   Typically, a tube that enables a measurement to be made or a sample to be obtained at a distance from the holder of the probe.

procarcinogen   A substance that is converted into a carcinogen as a result of its activation through the metabolic process.

process (OSHA)   Any activity or combination of activities, including use, storage, manufacturing, handling, and on-site movement of highly hazardous chemicals; or any group of vessels that are interconnected and separate vessels that are located such that a highly hazardous chemical (HHC) could be involved in a potential release (with some exceptions).

process flow diagram   A diagram or set of diagrams of a process depicting the equipment, flow of material, control schemes, raw materials, operating conditions including temperature and pressure, process stream characteristics, intermediates, and products.

process hazard analysis   [PHA]   A thorough, orderly, and systematic approach used to identify, evaluate, and control hazardous chemical processes. It involves a review of what could go wrong and what steps may be taken to safeguard against highly hazardous chemical releases.

process safety information   Written information on the highly hazardous chemicals, technology, and equipment associated with a process.

product liability   The liability of a manufacturer, processor, or nonmanufacturing seller arising from personal injury or property damage caused by a defective or dangerous product.

proficiency (Analytical)   The ability to satisfy a specified level of analytical performance.

Proficiency Analytical Testing program   [PAT program]   A program administered by the American Industrial Hygiene Association for evaluating the performance of industrial hygiene analytical laboratories and accrediting them if they meet specific requirements.

proficiency testing (Laboratory)   An interlaboratory testing program in which samples are sent to participating laboratories for analysis. The laboratory results are compared for the purpose of improving laboratory performance.

progeny   The decay products following the radioactive decay of a radionuclide.

prognosis   Outlook with regard to the outcome of an illness, such as complete recovery, partial recovery, or death.

proliferation   Reproducing or producing new growth or parts rapidly and repeatedly.

promoter   A substance, when administered at a later time, that enhances the tumorigenic response initiated by a primary or secondary carcinogen.

promulgate (Regulatory)   To put a law into effect through a formal public announcement, such as by publication in the Code of Federal Regulations.

propeller fan   A fan with airfoil blades that moves air in the general direction of the axis of the fan.

properties (Substance)   Characteristics such as the physical and chemical properties by which a substance can be identified.

prophylactic   A preventive treatment for the protection against a disease.

prophylaxis   Any substance or steps taken to prevent something from happening.

proportional counter   A gas-filled radiation detection device in which the signal produced is proportional to the number of ions formed in the gas by the primary ionizing particle.

prospective cohort study   A study that follows the outcome of interest (morbitity/mortality) a group (cohort) that is known to have been exposed to a condition/substance in the past or at present. The results are compared to the expected result as determined from a cohort of unexposed individuals.

prostrate   Lying down full length.

prostration   A state of total physical or mental exhaustion in which one becomes prostrate (i.e., lies down).

protection factor (Respiratory Protection) [PF] The ratio of the concentration of a contaminant in the ambient air to that inside a respirator.

protective clothing Special clothing that is worn to protect a worker from exposure to or contact with hazardous materials. Level A: Provides the highest level of protection from chemical hazards for both contact and respiratory protection. Level B: Provides the highest level of protection for respiration, but decreased skin protection. Level C: Provides minimum respiratory and skin protection and is used where there is no skin absorption hazard. Level D: No specific respiratory or skin protection is needed, and common work clothes are adequate.

protective hand cream A product designed to protect the hands from the harmful effects of some hazardous substances.

proteinuria The presence of an excess of serum proteins in the urine.

protocol The plan and/or procedures to be followed in a study or investigation.

proton Elementary nuclear particle with a positive electric charge equal numerically to the charge of the electron and a mass of 1.007277 mass units.

protoplasm A complex, colloidal substance conceived of as constituting the living matter of plant and animal cells, and performing the basic life functions.

prototype An original type, form, or instance that serves as a model on which later ones are based.

protoxicant The precursor of a toxic substance that is formed by the metabolic process.

prover tank A tank that is used to check the calibration of liquid flowmeters.

proximal Near the central part of the body, or hear a point of attachment or origin.

proximate cause The cause which produces the effect without the intervention of another cause.

pruritis Severe itching, usually of undamaged skin.

PSD (Air Pollution) Prevention of significant deterioration.

psi Pound(s) per square inch.

PSI Process safety information.

psia Pound(s) per square inch absolute.

psig Pound(s) per square inch gauge.

psittacosis A pneumonia-like viral disease that occurs in parrots and fowl, and that can be transmitted to man.

PSM Process safety management.

PSS TLV Particulate Size-Selective Threshold Limit Value.

psychogenic Originating in the mind or in mental activities and conditions.

psychogenic deafness Hearing loss due to a reaction to a physical or social environment. Also referred to as functional deafness.

psychometric testing The measurement of physiological variables using measures of memory, intelligence, aptitude, attention, personality, reaction time, and psychomotor function.

psychosis Mental disorder or severe emotional illness.

psychosocial factors (Indoor Air Quality)   Psychological, organizational, and personal stresses that may produce symptoms similar to those produced by poor indoor air quality.

psychosomatic   Pertaining to a partially or wholly psychogenic disease or physiologic disorder.

psychrometer   A device equipped with wet- and dry-bulb thermometers for determining the water vapor content of the air.

psychrometric chart   A graph showing the properties of moist air mixtures, such as relative humidity, dew point, etc., that can be used in air conditioning, ventilation, indoor air studies, and other applications.

PTFE   Polytetrafluoroethylene.

PTS   Passive tobacco smoke.

PTS   Permanent (hearing) threshold shift.

PTSD   Post-traumatic stress disorder.

Public Health Service   [PHS]   An arm of the Department of Health and Human Services that provides grants and loans for health care facilities, assists in the establishment and operation of emergency medical centers, addresses the occupational risks of federal employees, and is concerned with public health issues.

publicly owned treatment works   [POTW]   A device or system used in the treatment of municipal sewage or industrial wastes of a liquid nature, owned by a state or municipality.

pulmonary   Pertaining to the lungs.

pulmonary edema   Abnormal, diffuse, extravascular accumulation of fluid in the pulmonary tissues and air spaces of the lungs.

pulmonary emphysema   *See* emphysema.

pulmonary fibrosis   Progressive fibrosis of the pulmonary alveolar walls with steadily progressive difficult or labored breathing.

pulmonary function tests   Tests carried out to determine the capacity and health status of a person's lungs.

pulmonary hemosiderosis   Bleeding of the lungs that has been found in some cases to be a result of exposure to a mold.

pulse duration (Laser)   The length of time of an individual pulse of a laser output.

pumped sample   *See* active sampling.

pumping system (Lasers)   Method that is employed to raise the energy level of electrons in the lasing medium. They include optical, electrical, chemical, and others.

puncture wound   A wound resulting from the piercing through the skin by a sharp object or instrument.

pungent   Description of a taste or odor as strong or sharp.

pupil   The variable opening in the iris of the eye through which light travels into the interior of the eye.

pure tone (Acoustics)   A sound wave, the instantaneous sound pressure of which is a simple sinusoidal function of time; a sound wave characterized by its single frequency and whose wave form is that of a sine wave.

purging   The displacement of one material with another in process or other equipment.

purified protein derivative of tuberculin   A sterile, soluble, partially purified product of growth of the tubercle bacillus that is prepared in a special liquid medium free of protein and is used in the diagnosis of tuberculosis.

purpura   Disorders characterized by purplish or brownish-red discoloration of the skin, resulting from hemorrhage into the tissues.

purulent   Consisting of or containing pus.

pus   A liquid inflammation product made up of cells and a thin fluid.

push–pull hood   A hood consisting of an air supply system on one side of the contaminant source, blowing across the source and into a mechanical exhaust ventilated hood positioned on the opposite side.

pustule   A small elevation in the skin containing pus.

putrefaction   Partial decomposition of organic matter by microorganisms, producing foul-smelling matter.

putrescible   Subject to putrefaction, that is, the partial decomposition of organic matter by microorganisms, producing foul-smelling matter.

PVC   Polyvinyl chloride.

PWL (Acoustics)   Sound power level.

PWR   Pressurized water reactor.

pyrheliometer   Instrument for measuring the intensity of solar radiation.

pyrogen   A fever-producing agent of bacterial origin.

pyrolysis   The transformation of a compound into one or more other substances by heat alone. A process by which a material is decomposed by heating it in the absence of air; thus it is similar to destructive distillation.

pyrometer   An instrument for measuring or recording temperature above the range of a mercury thermometer.

pyrophoric material   A material that will spontaneously ignite when exposed in dry or moist air below 130 degrees F.

Q   Quantity or volume of air.

QA   Quality assurance.

QC   Quality control.

QIPs   Quality Improvement Program as detailed in the Clean Air Act Amendments.

QLFT   Qualitative fit test.

qlty   Quality.

QNFT   Quantitative fit test.

qs   As much as sufficient.

qt   Quart.

q.t.   In secret.

qty   Quantity.

quadrant   One-fourth of the circumference of a circle.

qualified (OSHA)   The characteristic of one who, by possession of a recognized degree, certificate, or professional standing, or by extensive knowledge, training, and experience, has successfully demonstrated his/her ability to solve or resolve problems relating to the subject matter, the work, or the project.

qualitative   The characteristic attributes or qualities pertaining to an exposure based on subjective information, nonrigorous quantitative data, and judgment. Descriptive and nonnumerical.

qualitative analysis (Chemistry)   The chemical determination of the constituents of a substance without regard to quantity.

qualitative exposure assessment   The identification of contaminants and physical agents to which an individual may be exposed, and a judgment of the associated hazard based on the frequency and duration of exposure, the control measures in effect (engineering, administrative, and personal protection), the properties of the stressor, and the manner in which it is being used/handled.

qualitative fit test   [QLFT]   A pass/fail test to assure the adequacy of respirator fit that relies on the individual's response to a test agent.

quality   A characteristic or an attribute of something.

quality assurance   [QA]   A system of practices, procedures, and activities that are taken to provide assurance that the work being carried out will meet defined standards of quality. The assessment of the potential for a procedure to produce sampling results of adequate quality to satisfy the defined objectives. The primary purpose of a quality assurance program is to provide the necessary safeguards to minimize erroneous sample analyses and to provide a means of detecting errors when they occur.

quality control   [QC]   The specific procedures that are to be adhered to in order to ensure that the sampling and analytical method results meet defined objectives.

quality factor (Ionizing Radiation)   [QF]   A modifying factor that is used to derive the radiation dose equivalent from absorbed dose. It is a factor by which absorbed radiation dose in rad is multiplied to obtain a quantity that expresses the biological effectiveness of the absorbed dose in rem. The factor for beta, gamma, and X-radiation is 1. For alpha particles and fast neutrons, it is 10. Other values are used for neutrons of other energies and heavy recoil nuclei.

quality of life   The degree to which persons perceive themselves able to function physically, emotionally, and socially.

quantify   To determine or express the quantity or amount of something.

quantitative   The property of anything that can be determined by measurement and expressed as a quantity.

quantitative analysis (Chemistry)   The chemical determination of the amounts or proportions of constituents in a substance.

quantitative exposure assessment   The procedure of quantitatively determining an individual's exposure to a health hazard, employing accepted sampling and analytical procedures, and assessing the likelihood that an adverse health effect may occur based on the monitoring results.

quantitative fit test   [QNFT]   An assessment of the adequacy of respirator fit by numerically measuring the amount of leakage into the respirator.

quantum   The smallest indivisible quantity of radiant energy. Also referred to as a photon.

quantum theory   A concept that energy is radiated intermittently in units of definite magnitude called quanta, and is absorbed in the same manner.

quarantine   The limitation of freedom of movement of well persons or domestic animals that have been exposed to a communicable disease, for a period of time equal to the longest incubation period of the disease, in such manner as to prevent effective contact with those not exposed.

quart   [qt]   Volume measurement equivalent to 946 millimeters.

quartile   The value of the boundary at the 25th, 50th, or 75th percentile of a frequency distribution divided into four parts, each containing one-quarter of the population.

quartz   One of the forms of crystalline silicon dioxide. Also referred to as one of the forms of free silica.

quasi   Almost or somewhat.

quaternary ammonium compounds   [Quats]   Odorless, noncorrosive, relatively nontoxic, and stable cationic detergent compounds that have bacteriostatic, tuberculostatic, sporostatic, fungistatic, and algistatic properties at low concentrations.

quell   Suppress.

quick-drench shower   A facility, positioned in the immediate work area where there is a possibility of a hazard to the skin, such as corrosion, skin absorption, etc., for quickly drenching the body with water.

quicksilver   Mercury.

quid pro quo   An equal exchange of something for something.

quiescent   Inactive or dormant.

quorum   The minimum number of members of a committee or organization that must be present for the valid transaction of business.

R

R   Degrees Rankine.

RACT   Reasonably achievable control technology.

rad   Radiation absorbed dose.

radian   The angular measure equal to the angle subtended at the center of a circle by a chord equal in length to the radius of the circle.

radiant energy   Any of the forms of radiant energy (e.g., heat, light, electromagnetic waves, ionizing radiation, etc.) radiating from a source. Energy that travels outward in all direction from its source.

radiant flux   The rate of energy incident on a specified area of a material, such as watts per square centimeter.

radiant heat   A form of electromagnetic energy.

radiant heat temperature   The temperature of an object as a result of its having absorbed radiant energy.

radiant heat transfer   The transfer of heat energy in wave form from a hot object to a colder object.

radiation   The emission and propagation of energy through space or through a material medium in the form of waves in all directions from a center or focal point; for instance, the emission and propagation of electromagnetic waves, electric waves, or other forms of electromagnetic radiation, as well as ionizing radiation.

radiation absorbed dose   [rad]   Special unit of absorbed ionizing radiation dose equal to an absorbed dose equivalent of 100 ergs per gram of material or 0.01 joules per kilogram (0.01 gray). A measure of absorbed dose to body tissue.

radiation area   An area accessible to individuals in which ionizing radiation levels could result in a person receiving a dose equivalent in excess of 5 millirem (equivalent to 0.05 millisieverts) in 1 hour at 12 inches (30 centimeters) from the source or from any surface that the radiation penetrates, or a dose in excess of 100 millirem in any 5 consecutive days.

radiation illness   Illness resulting from exposure to ionizing radiation, including general malaise, nausea, vomiting, and diarrhea. Also referred to as radiation sickness.

radiation protection guide   [RPG]   The ionizing radiation dose that should not be exceeded without careful consideration of the reasons for doing so and consideration of the person(s) receiving the dose.

radiation protection officer   [RPO]   The person who has been selected and trained to be responsible for overseeing the ionizing radiation protection program in a facility. Also referred as the radiation safety officer.

radiation safety officer   [RSO]   *See* radiation protection officer.

radiation shielding   The reduction of the radiation level by interposing a shield of absorbing material between a radioactive source and a person, work area, or radiation-sensitive device.

radiation sickness   Sickness that can be fatal that results from receiving a large dose of ionizing radiation within a short period of time.

radiation source (Ionizing Radiation)   A naturally occurring or man-made source of ionizing radiation, such as is in a device or piece of equipment that emits or, when energized, produces ionizing radiation.

radiation survey (Ionizing Radiation)   An evaluation of the radiation hazard incident to the production, use, release, disposal, or presence of a radioactive material or other source of ionizing radiation under a specific set of conditions. Includes surveys necessary to evaluate surface contamination, external exposures to personnel, and the concentration of airborne radioactive materials in the facility and in effluents from the facility, as appropriate.

radical (Chemistry)   An ionic group having one or more charges, either positive or negative. A group of atoms that can enter into a chemical reaction, but that are incapable of existing separately.

radioactive   A property of some materials/elements that is characterized by their spontaneously emitting ionizing radiations.

radioactive contamination   The deposition of radioactive material in any place where it may harm persons or equipment.

radioactive decay   The disintegration of the nucleus of an unstable nuclide by the spontaneous emission of charged particles and/or photons.

radioactive material   A naturally occurring or artificially produced substance that is a solid, liquid, or gas, and that emits ionizing radiation spontaneously.

radioactive material label   A label indicating "Caution—Radioactive Material" on any container in which is transported, stored, or used, a quantity of a radioactive material greater than is specified for that radionuclide.

radioactive series   A succession of nuclides, each of which transforms by radioactive disintegration into the next nuclide until a stable one results.

radioactive waste   Waste that contains materials that are radioactive and that must be disposed of according to regulatory requirements.

radioactivity   The property of certain nuclides to spontaneously emit particles or gamma radiation, or to emit X-radiation following orbital electron capture, or to undergo spontaneous fission.

radiobiology   The branch of biology which deals with the effects of ionizing radiation on biological systems.

radiochemical   Any compound or mixture containing sufficient radioactive elements to be detected by a geiger counter.

radiochemistry   Chemistry concerned with the properties and behavior of radioactive materials.

radio frequency   The frequency range from 300 kilohertz to 100 gigahertz.

radiograph   A picture of an object that is made by passing ionizing radiation through the object onto photographic film positioned on the opposite side of the object from the radiation source.

radiographer   The individual who is in attendance at a site where ionizing radiation sources are being used and uses or supervises their use in industrial radiographic operations. This individual is responsible for ensuring the operation is in compliance with regulations and in adherence with good practice during the procedure.

radiography   The use of penetrating ionizing radiation to make photographs of the inside of objects. An examination of humans or animals, or of the structure of materials by nondestructive methods, utilizing sealed sources of ionizing radiation or ionizing radiation-producing machines.

radioisotope   An unstable isotope of an element that disintegrates spontaneously, emitting ionizing radiation and yielding a different isotope.

radio-labeled   A chemical to which a radioactive tracer element has been added.

radiological health   The art and science of protecting humans, animals, and the environment from injury or damage from sources of ionizing radiation, and of promoting better health through beneficial applications of sources of ionizing radiation.

radiological work permit   [RWP]   A work permit that identifies work conditions, site worker protection measures, and monitoring requirements, and establishes administrative controls to be adhered to.

radiology   A branch of medicine that deals with the diagnostic and therapeutic applications of X-rays and radioisotopes.

radiomimetic   Imitating the biological effects of ionizing radiation.

radionuclide   Any naturally occurring or artificially produced radioactive element or isotope.

radiosensitive   Term used in describing tissues that are more easily damaged as the result of exposure to ionizing radiation.

radiosonde   Instrumentation carried up into the air by a balloon to measure temperature, pressure, and humidity at various heights in the atmosphere.

radiotherapy   Treatment of ailments by the application of doses of ionizing radiation from various sources.

radiotoxicity   The ability of a radionuclide to produce a local or systemic toxic effect as a result of the radiation emitted from the material.

radon   Radioactive gas produced by the decay of radium 226 or radium 224.

radon decay products   *See* radon progeny.

radon daughters   *See* radon progeny.

radon progeny   Collectively, the intermediate products produced in the radon decay chain. Also called radon daughters or radon decay products.

raffinate   In solvent extraction, that portion of the mixture which remains undissolved and not removed by the solvent.

rafter sample   A sample of settled dust that is obtained from a rafter or other undisturbed surface that will contain representative particulates that have settled out of the air. The sample must be representative of the airborne dust to which personnel are exposed.

ragweed   A variety of weeds that produce profuse amounts of pollen that are a chief cause of hay fever.

rainbow passage   A paragraph of text that, when read, results in the reader making a wide range of facial movements. This reading can be used for the talking phase of the respirator fit test protocol.

rain cap   A sheet metal fixture that is placed on the outlet of a stack/vent for preventing rain from entering. Also called a weather cap.

rales   Abnormal sounds in the respiratory system indicating some type of pathological condition.

R & D   Research and development.

random   Not deterministic. A variable whose value at a particular future instant cannot be predicted exactly.

random errors   Errors originating in uncontrollable or unknown sources, the result of variation due to chance that occurs in monitoring despite the effort to control all variables. They are characterized by the random occurrence of both positive and negative deviations from the mean, and they tend to cancel out if the sample size is sufficient.

random noise   An oscillation whose magnitude is not specified and cannot be predicted with certainty for any given instant of time.

random sample   A sample that has been collected in such a manner that each individual in the population represented by the sampled individual had an equal probability of being sampled. This concept can be applied to personnel, work areas, work shifts, dates, etc. The objective in collecting a random sample is to obtain a sample which is free of bias.

range (Data)   A measure of the variability in a data set; the difference between the highest and lowest values in the set.

range (Instrument)   The upper and lower limits between which an instrument responds and over which the instrument is calibrated. The interval between the upper and lower measuring limits of an instrument.

Rankine   [R]   A temperature scale with 0°F at 460°R. The freezing point of water on this scale is 491.6°R, and the boiling point is 671.7°R.

RAPF   Recommended assigned protection factor.

rash   A temporary cutaneous eruption, reddish coloring, or blotch on some part of the skin.

rate   A quantitative measure of a part of the whole, such as a death rate of 56 persons in a total population of 5000.

rate of decay (Acoustics)   The time rate at which the sound pressure level decreases at a given point and at a given time after the source is turned off.

Raynaud's disease   *See* Raynaud's syndrome.

Raynaud's syndrome   A vascular disorder resulting in the constriction of the blood vessels of the hands due to cold temperature, emotions, or unknown cause. The hands become a bluish-white color due to lack of blood circulation and become painful upon exposure to cold. Also referred to as Raynaud's disease, dead hands, and vibration white-hands disease.

RBC   Red blood cell.

RCF   Refractory ceramic fiber.

RCRA   Resource Conservation and Recovery Act.

reaction (Physiology)   The response of an organism to a stimulus.

reaction (Chemical)   A process in which a chemical change takes place, such as one or more substances being transformed into other substances.

reaction time   The time required for a person to react to a stimulus.

reactive gases   Gases that will react with other materials or within themselves by a reaction other than burning, with the production of potentially hazardous quantities of heat or reaction products.

reactive material   A material that by itself is readily capable of detonation or explosive decomposition, or explosive reaction at normal or elevated temperature and pressure.

reactive muffler (Acoustics)   A type of muffler used to reduce noise emissions from an engine, such as that used in an automobile exhaust system.

reactivity   A measure of the tendency of a substance to undergo chemical reaction with the release of energy; the susceptibility of material to release energy.

readout (Instrument)   The indication on the meter (or readout of an instrument) when exposed to the contaminant or stressor being measured.

reagent   A substance used in a chemical reaction to produce another substance or for the detection, measurement, or analysis of other materials.

reagent blank   Materials used in sample analysis are evaluated as reagent blanks to determine their contribution, if any, to the analytical result.

real time (Instrument)   An instrument that responds to and indicates a contaminant concentration or level of a physical agent as changes are occurring, thus indicating results that exist at the time of measurement.

reasonably achievable control technology   [RACT]   Emission controls required on existing sources in non-attainment areas under the Clean Air Act.

rebuttal   Response to a statement, such as opposing, contradicting, or refuting it.

receptor   A structure that receives information or responds to stimuli.

recidivism   Reverting to a previous condition, pattern, or behavioral mode.

recirculated air   *See* recirculation.

recirculation (Ventilation)   Air withdrawn from a space, passed through a ventilation system, and delivered again to an occupied space.

reclamation   The restoration of land, water, or waste materials to usefulness through methods such as sanitary landfill, wastewater treatment, or material recovery.

recognizable hazard (OSHA)   A hazardous condition that is recognized as a hazard; that is, a hazard of which the employer had actual knowledge of the hazard or one recognized as a hazard within the employer's industry.

recommended exposure limit   [REL]   An occupational exposure limit recommended by NIOSH as being protective of worker health over a working lifetime.

record (OSHA)   Any item, collection, or grouping of information, regardless of the form or process by which it is maintained.

recoverable resources   Materials that still have useful physical, chemical, or biological properties after serving their original purpose and that can be reused or recycled for the same or other purposes.

recovery efficiency   The ratio, expressed as a percentage, of the amount of a material recovered from a sampling media to the amount placed on/in the media.

rectifier   A device for converting alternating current to direct current.

recuperate   To recover from an injury or illness.

recycled material   A material that is used in place of a raw virgin material in the manufacture of products and that consists of consumer waste, industrial scrap, materials from agricultural product waste, and other materials.

recycling   The procedure whereby waste materials are reused for the manufacture on new materials and goods.

red blood cell   Erythrocyte.

red blood count   The number of erythrocytes in one milliliter of blood.

reduction (Chemistry)   The addition of one or more electrons to an atom through chemical change.

redundancy   The providing of devices to duplicate the function of others in the event that one fails.

reentrainment   Situation that occurs when the air that is exhausted from a building is brought back into the building through an air intake or other opening in the building.

reentry   *See* reentrainment.

reference man   A hypothetical aggregation of human physical and physiological characteristics arrived at by international consensus. For example, the weight, height, and other physical dimensions presented for what has been agreed to as the reference man. Also referred to as the standard man.

reference method (Sampling/Analysis)   A sampling and analytical method that has been validated and is recommended for determining the concentration of an air contaminant to which workers (industrial hygiene) or the general population in the community (environmental) may be exposed.

reflectance   A measure of the ratio of the luminance of a surface to the illumination on the surface.

reflex   An involuntary response to a stimulus.

reformulated gasoline   Gasoline whose composition has been altered in order to reduce evaporation and exhaust emissions that contribute to ozone formation.

refraction (Acoustics)   The bending of a sound wave from its original path due to its passing from one medium to another, or due to a temperature or wind gradient.

refractive index   The ratio of the velocity of light at a specific wavelength in air to its velocity in a substance under examination.

refrigerant   A substance that will absorb heat while vaporizing and whose boiling point and other properties make it useful as a medium for refrigeration.

refuse   Anything discarded as useless or worthless trash.

register   A fixture through which air is returned to a ventilation system. Also referred to as a return air register.

registered trademark   A trademark that is registered with the U.S. Patent and Trademark Office, thereby securing its exclusive use to the person registering it.

Registry of Toxic Effects of Chemical Substances   A NIOSH Publication containing information on the acute and chronic toxicity of potentially toxic chemicals, as well as their exposure limits and status under various federal regulations and programs. Accessible in an on-line interactive version.

regression   A statistical procedure that is employed to establish a relationship between two variables to enable the prediction of the values of one variable, Y (dependent variable), to those which correspond to given values of the other variable, X (independent variable).

regression analysis   *See* regression.

regression line   A straight line that summarizes the relationship between two variables.

regulated area (OSHA)   An area where exposure to a regulated airborne contaminant or physical stress agent is or can be expected to be in excess of an OSHA permissible exposure limit.

regulated waste (Bloodborne Pathogen)   Waste that includes liquid or semi-liquid blood or other potentially infectious material (OPIM); items contaminated with blood or OPIM that would release these substances in a liquid of semi-liquid state if compressed; items caked with dried blood OPIM that are capable of releasing these materials during handling; contaminated sharps; and pathological or microbiological wastes containing blood or OPIM.

regulator   Device for reducing the pressure or flow of liquid in a line.

regurgitation   The return of solids or fluids to the mouth from the stomach.

rehabilitation (Housing)   The elimination of those properties that cannot be brought up to standard (due to, for example, lead, asbestos, location, etc.) and the rebuilding of those that can be brought up to standard.

Reid vapor pressure   [RVP]   The vapor pressure of a liquid at 200°F, as determined by a standard laboratory procedure (ASTM Test D-23) and expressed in pounds per square inch absolute.

reinfestation   The return of rodents, insects, or other pests after an extermination has been carried out.

REL   Recommended exposure limit (set by NIOSH).

relapse   To recur, such as the recurrence of a disease or symptom.

relative humidity   [RH]   The ratio of the actual partial pressure of the water vapor in a space to the saturation pressure of pure water at the same temperature. The ratio of water vapor at a given temperature to the vapor pressure corresponding to saturation at that temperature.

relative risk   A measure of the association between exposure to a factor and the occurrence of a disease expressed as a ratio of the incidence rate in exposed persons to the incidence rate in unexposed persons.

reliability (General)   The repeatability of functioning free from failure or from the degradation of performance.

reliability (Instrument)   The ability of an instrument and its components to retain their operating performance characteristics over a reasonable period of use. A statistical term having to do with the probability that an instrument's repeatability and accuracy will continue to fall within specified limits. This is a very important characteristic for instruments that are to be used in field applications.

relief   The reduction or removal of pain or suffering.

relief valve   A safety valve actuated by inlet pressure that opens relative to the increase in pressure above its opening pressure.

rem   A special unit of ionizing radiation dose equivalent, the initials of which stand for roentgen equivalent man. The unit of dose of ionizing radiation that produces the same biological effect as a unit of absorbed dose of X-rays. Enables one to compare the dose from one type of ionizing radiation to another.

remedial   Acting as a remedy.

remediate   To remedy, correct, or resolve a problem.

remedy   Any agent or activity that relieves or cures the appearance or symptoms of an illness.

remission   Abatement or reduction of the symptoms or signs of an illness.

renal   Related to or associated with the kidney.

renal failure   The failure of the kidney(s) to maintain water and electrolyte balance in the body.

repeatability (Instrument)   The ability of an instrument to reproduce readings repeatedly when sampling the same concentration.

repeatability (Sampling)   The closeness of agreement between samples that are collected simultaneously.

repeat violation (OSHA)   A violation of a standard, regulation, rule, or order in which, upon reinspection, a substantially similar or the same violation is found.

repeat violation citation (OSHA)   A citation that is issued when the original violation of a standard has been abated, but upon reinspection, another violation of the previously cited section of the standard is found.

repellent   A chemical applied to the skin, clothing, or other location, to discourage arthropods from landing on and attacking individuals, or to discourage other agents, such as larvae, from penetrating the skin.

repetitive stress injury   [RSI]   Musculoskeletal disorder resulting from repeated trauma.

replacement air   Air provided to a space to replace air that is being exhausted. Also referred to as make-up air.

replicate samples   More than one sample collected at the same time and place for the purpose of determining their reproducibility.

replication   The process or act of duplicating or reproducing something, such as an experiment or test, and obtaining the same findings. Carrying out an experiment or analytical procedure to confirm the findings, increase precision, and/or obtain a closer estimation of the error.

reportable quantity [RQ]   The quantity of a hazardous substance that is considered reportable under CERCLA.

representative sample (General)   A large enough sample of a universe that can reasonably be expected to exhibit the average properties of that universe.

representative sample (Industrial Hygiene)   A sample obtained as being representative of the exposure of an individual to a hazardous substance or physical agent during the work activity being performed.

reproducibility (Chemical Analysis)   A measure of the deviation of test results from their mean value.

reproducibility (Instrument)   The precision of a single measurement on the same sample made by different operators using different instruments of the same type.

reproductive disorders   Disorders that include low sperm count, infertility, impotence, genital deformities, cancers of the breast and prostate, endometriosis, menstrual disorders, spontaneous abortions, low birth weight, preterm births, birth defects, developmental disabilities, and neurological disorders.

reproductive toxicity   A harmful effect to the adult reproductive system. The ability of a substance or physical agent to adversely affect the reproductive system.

reproductive toxin   A substance that has the capability to adversely affect the adult reproductive system of either male or female.

research octane number [RON]   Measure employing ASTM methods of the knock characteristics of a gasoline product.

reservoir of infection   Man, animals, plants, soil, or organic matter in which an infectious agent lives and multiplies and depends on for survival, reproducing itself in such a manner that it can be transmitted to man.

residence time   The length of time a substance remains in the body.

residual fuel   A heavy oil product that is used by utilities and other industry as a fuel.

residual volume (Spirometry) [RV]   The amount of air remaining in the lungs following a maximum expiration.

resilient   The ability of a material to recover quickly and assume its original shape following bending, compression, or other change.

resistance   The sum total of body mechanisms that interpose barriers to the invasion or multiplication of infectious agents or to damage their toxic products.

resolution (Instrument)   The smallest change in concentration of a contaminant that will produce a detectable change in instrument output.

resolving power (Microscope)   The smallest distance by which two points can be separated and still be observed as two separate points by a given lens.

resonance (Acoustics)   A state that exists when any change, however small, in the frequency of excitation causes a decrease in the response of the system.

resonant frequency (Acoustics)   A frequency at which resonance exists.

Resource Conservation and Recovery Act [RCRA]   Federal law that defines the requirements for the safe transport, storage, and disposal of hazardous wastes.

resource recovery The recovery of materials or energy from solid waste.

respirable Suitable for respiration, such as aerosols of a size small enough to be inhaled into the deep lung space (i.e., particulate matter with an aerodynamic diameter of 10 micrometers or less).

respirable dust *See* respirable particulates.

respirable fraction The mass fraction of inhaled particulate matter that penetrates to the unciliated airways of the lungs.

respirable particulate The fraction of inspired particulates that are capable of penetrating into the lung after larger particles are removed in the upper respiratory tract. Respirable particulates are those in the size range that permits them to penetrate to the lungs upon inhalation.

Respirable Particulate Mass (Sampling) Those particles that penetrate a separator whose size collection efficiency is described by a cumulative lognormal function with a median aerodynamic diameter of 3.5 micrometers and with a geometric standard deviation of 1.5.

Respirable Particulate Mass TLVs [RPM TLVs] Those materials that are hazardous when deposited in the gas-exchange region of the respiratory system.

respirator A personal protective device that is designed to protect the wearer from inhaling a harmful atmosphere. There are two basic types of respirators—one that removes the contaminant from inspired air (air-purifying), and one that supplies clean air from another source, such as a cylinder or compressor (atmosphere-supplying).

respirator fit test A procedure that is followed to determine if a respirator wearer obtains a proper fit, either by a qualitative, quantitative, or workplace test. The results of the test indicate whether or not the wearer is getting the protection that is to be afforded by the respiratory protection device and whether the user puts the device on properly to get a good fit.

respirator use (OSHA) Use of respirators in those cases in which engineering controls are not feasible, have not yet been installed, or are not adequate to reduce exposures below a permissible exposure limit.

respiratory diseases Diseases that result from the effects of harmful substances on the respiratory tract (e.g., pneumoconiosis, bronchitis, pneumonitis, pulmonary irritation, lung cancer, etc.).

respiratory irritants Substances that irritate the respiratory tract (e.g., the nasal passages, larynx, trachea, bronchi, alveoli, etc.).

respiratory protection An apparatus, such as a respirator, used to reduce the individual's intake of a substance by the inhalation route.

respiratory protection program Typically a written program addressing the procedures for the selection, training, inspection, maintenance, storage, use, etc. necessary to have an effective respirator program for protecting personnel from inhalation hazards when engineering or administrative controls are not adequate or are not being implemented, or for tasks that are intermittent and infeasible for engineering control.

respiratory rate The number of inhalations/exhalations per unit of time, such as per minute.

respiratory system Consists of, in descending order, the nose, mouth, nasal passages, nasal pharynx, pharynx, larynx, trachea, bronchi, bronchioles, air sacs (alveoli) of the lungs, and the muscles of respiration.

respire   To breathe.

response (Instrument)   The quantity of output signal that results from a challenge by a given amount of sample (i.e., the material of interest).

response (Physiology)   The activity of an organism or individual that follows as a result of a stimulus.

response check (Instrument)   A procedure to determine that the instrument is working properly and responding to the contaminant it was designed to detect and measure. Typically, no adjustment is made to the instrument when a response check is made.

response time (Instrument)   The time required for an instrument to indicate a designated percentage (usually 90 percent) of a step change in the variable being measured. The time required for an instrument to indicate a change in concentration or level after being challenged by the agent being measured.

responsibility   Having to answer for activities and results.

restricted area (Ionizing Radiation)   An area, access to which is limited by the licensee for the purpose of protecting individuals against undue risks from exposure to ionizing radiation and radioactive materials.

restricted duty   When an injured employee is unable to do the normal full scope of the job or must be transferred to another job temporarily because of his restrictions, but is able to come to work.

retching   An involuntary attempt at vomiting.

retention   The amount of a substance deposited in the body less the amount excreted/exhaled.

retention time (Gas Chromatography)   The time period from the moment of injection of a sample to the moment the substance of interest elutes from the column and is detected by the detector.

reticle   A glass disc with a scale inscribed on its surface that is placed in the eyepiece of a microscope to define an area and determine the size of particles.

retina   The delicate multilayer, light-sensitive membrane lining the inner eyeball and connected by the optic nerve to the brain.

retinitis   Inflammation of the retina.

retrospective cohort study   A group (cohort) that is known to have been exposed to a condition/substance in the past that is selected and followed to disease or death at some point also in the past. The results are compared to the expected number of occurrences found in an unexposed cohort from the same time period.

return air   Air returning to a heater or air conditioner from a heated or air conditioned space for conditioning and recirculation.

reverberant field (Acoustics)   Location where reflected sound dominates as opposed to a location near the source where direct sound from the source dominates.

reverberant room   A room with hard walls, floor, and ceiling, such that sound is scattered and reflected, and persists for a short period after the noise source within the room is turned off.

reverberation   The persistence of sound in an enclosed space as a result of multiple reflections after a source has stopped emitting sound energy.

reverberation time (Acoustics) Time required for the mean squared sound pressure level, originally in a steady state, to decrease 60 decibels after the source of noise has been terminated.

reverse osmosis [RO] A process of forcing water through a semipermeable membrane that allows the passage of the water but not of other materials, e.g. solutes.

Reynolds number [Re number] A dimensionless ratio applicable to the movement of fluid through a pipe/duct that is proportional to pipe or duct diameter and the velocity and density of the fluid, and inversely proportional to its viscosity. The Reynolds number is used to predict whether fluid flow is turbulent or laminar.

rf Radio frequency.

RFG Reformulated gasoline.

RFI Request for information.

R.F.I. Radio frequency interference.

RFR Radio frequency radiation.

rf radiation Radio frequency radiation.

RH Relative humidity.

rheology The study of the deformation and flow of matter.

rhinitis Inflammation of the mucous membrane lining of the nasal passages.

rigging Construction in a makeshift manner.

rigidity Tenseness or inflexibility.

rigor mortis The stiffness of the joints and body of a dead body.

ring badge A film badge or TLD that is worn on the finger to determine the wearer's exposure to ionizing radiation.

Ringelmann Chart Chart, numbered from one to five, that simulates various smoke densities by presenting different percentages of black. Used to evaluate the emission of smoke from a stack.

riparian rights The entitlement of a land owner to the water on or bordering his/her property, including the right to prevent diversion or misuse of it upstream.

rise time (Instrument) The time required for an instrument to indicate a designated percentage (e.g., usually 90 percent) of the full response that will result with an increase in the concentration of the material being measured.

risk A measure of the probability and severity of an adverse health effect occurring as a result of an exposure to a contaminant, physical stress, or other health hazard (e.g., ergonomic factor, biological organism, bloodborne pathogen, etc.).

risk assessment A qualitative or quantitiative evaluation of the likelihood that an adverse health effect may occur under the prevalent conditions or others that are likely to develop. Factors to consider in a qualitative risk assessment are the toxicity of the material, the frequency and duration of exposure/contact, control measures (engineering, administrative, or personal protective equipment) in use and their effectiveness, the properties of the material, and conditions of use (e.g., temperature, pressure, volume, etc.).

risk-benefit analysis   The process of analyzing and comparing, on a single scale, the expected positive (i.e., benefits) and negative (i.e., risks/costs) results of an action.

risk factors (General)   Factors that predispose individuals to greater risk, such as the characteristics of the exposure group, including weight, sex, race, age, health status, contaminants and level of exposure, etc., and that may be associated with an increase in the probability of a toxic effect occurring.

risk factor (Toxicology)   The excess risk per unit of dose at a specified dose level.

risk group   The actual or hypothetical exposure group composed of the general population, workers, etc.

risk management   The complex judgment and analysis that uses the results of risk assessments to produce decisions about environmental actions to be initiated.

RMP   Risk management program or risk management plan.

RMP Program   Radon Measurement Proficiency Program.

rms   Root mean square.

RN   Registered nurse.

ROC   Reactive organic compound.

rodenticide   A material used to destroy rodents, generally through ingestion, or to prevent them from damaging foodstuff.

roentgen   [R]   A unit of exposure to ionizing radiation. The amount of gamma or X-radiation required to produce ions carrying 1 electrostatic unit of electrical charge in 1 cubic centimeter of dry air at standard conditions.

roentgen equivalent man   [rem]   The unit of dose of any ionizing radiation that produces the same biological effect as a unit of absorbed dose of ordinary X-rays. The dose equivalent in roentgen equivalent man is equal to the absorbed dose in grays multiplied by an appropriate quality factor.

room air balance   A general term describing the pressure differentials and desired flow direction of room air with respect to adjacent spaces and from room to room within a structure.

root mean square   [rms]   Square root of the arithmetic mean of the squares of a set of values of a function of time or other variable.

rotameter   A flow metering device consisting of a precision bored, tapered, transparent tube with a solid float inside. With air flowing through the device, the float rises inside the tube until the pressure drop across the anular area between the float and tube wall is just sufficient to support the float. A rotameter is considered a secondary calibration standard and must be calibrated against a primary standard to obtain accurate results.

rotating vane anemometer   A device for measuring air velocity.

route of entry   *See* routes of exposure.

routes of exposure   The avenues by which toxic substances enter the body. The most common routes for industrial exposure are inhalation and dermal absorption (skin absorption) for liquid or solid materials. A less important industrial route of exposure is ingestion. Punctures are a route of exposure for some materials (e.g., radioactive materials). Also referred to as routes of entry.

routine monitoring   Involves the frequent and regular industrial hygiene sampling for determining employees' exposure to a substance to which personnel are somewhat routinely exposed or with which they work frequently.

routine respirator use   The wearing of a respiratory protective device while carrying out a regular and frequently repeated task.

RPG   Radiation Protection Guide.

rpm   Revolutions per minute.

RPM TLVs   Respirable Particulate Mass TLVs.

RPO   Radiation protection officer.

RQ   Reportable quantity.

RSI   Repetitive strain injury.

RSO   Radiation safety officer.

RTECS   Registry of Toxic Effects of Chemical Substances.

rubefacient   Substance that is able to redden the skin.

runoff   Precipitation that flows over the ground and does not soak in, but flows into a stream or other body of water.

rupture disk   The operating part of a pressure relief device that, when installed in the device, is designed to rupture at a predetermined pressure and permit discharge of the contents.

rust   Coating on iron from corrosion as a result of the action of oxygen and water to form hydrated iron oxide.

RVP   Reid vapor pressure.

RWP   Radiological work permit.

S

s   Second.

Sabin   A unit of measure of sound absorption.

saccharin   A very sweet tasting powder that can be used in qualitative fit testing of particulate-type air-purifying (filter) respirators.

SAE   Society of Automotive Engineers.

safe   A condition or situation that is reasonably free from hazards that may cause injury or adverse health effects.

safe day   A workday in which there were no lost time injuries.

safety   The proper handling of a substance or carrying out of a task to eliminate its capacity to cause injury or do harm.

safety belt   Device worn around the waist or as a harness for securing a person to a structure or within a vehicle.

safety can   An approved container of not more than 5 gallons capacity, having a spring-closing lid and spout cover, and so designed that it will safely relieve pressure when subjected to fire exposure.

safety engineering   Discipline concerned with the planning, development, implementation, maintenance, and evaluation of the safety aspects of equipment, the environment, procedures, operations, and systems to achieve effective protection of people and property.

safety factor   The ratio of the normal working condition to the ultimate condition.

safety guard (Grinder)   Enclosure for containing pieces of the grinding wheel in the event the wheel is broken during use.

safety professional   An individual with specialized skills, knowledge, and/or education who has achieved professional status in the safety profession.

safety relief valve   A valve fitted on a pressure vessel or other containment under pressure to relieve overpressure.

safety valve   A pressure relief valve that is actuated by inlet pressure and with resultant rapid opening above its pressure setting.

salamander   Portable type of furnace without grates that is used as a space heater in some work locations.

salinity   The degree of salt in the water.

saliva   The clear, alkaline secretions from the mucous glands of the mouth.

salivation   The excessive excretion of saliva.

sample (Industrial Hygiene)   An airborne contaminant (fume, dust, mist, vapor, etc.) collected from the air within the workplace, or the measurement of the level of a physical agent (noise, heat, ionizing radiation, etc.) to which workers are exposed. The sample must be random and representative of the exposure of the individual sampled, as well as collected in an acceptable manner so that it can be compared to an established exposure standard.

sample (Statistics)   The part or subset of a population that is selected for statistical analysis.

sample blank   The gross instrument response attributable to reagents, solvents, and the sample media used in air sampling and subsequent analysis of samples. *See* blank.

sample draw (Sampling)   The procedure and method used to cause the deliberate flow of the atmosphere being monitored to a sensing element. *See* active sampling.

sample parameters   Estimators of population parameters, such as the mean, standard deviation, etc., based on observations of a subset of the population.

sample storage stability   The period of time, in days, over which storage losses of analytes are generally less than 10 percent, provided that storage and shipment precautions are observed. It is determined by collecting a number of samples at the level of concern (e.g., the TLV) at room temperature and about 80 percent relative humidity, and subsequently analyzing sets of these samples (e.g., about 6 in each set) over a 2-week or longer period to determine losses that may occur during storage.

sampling and analytical method bias   An estimate of accuracy for the sampling and analytical method as determined by sampling a test atmosphere and analyzing the sampling media. The net bias for a given concentration is the difference between the true test atmospheric level and the sampling method concentration, expressed as a percentage of the true test atmosphere concentration.

sampling frequency   The time interval between the collection of successive samples.

sampling media   *See* media.

sampling period   The length of time over which a sample is collected.

sandblasting   A method for cleaning surfaces employing sand as the abrasive material. Also used in a generic sense for abrasive cleaning operations.

sandhog   Worker engaged in tunneling work in which atmospheric pressure control is required.

sanitarian   A scientist and public health educator who uses the knowledge and skills of the natural and environmental sciences to prevent disease and injury, and to promote well being.

sanitary landfill   A facility for the disposal of solid wastes employing an engineered method of disposal on land in a manner that minimizes environmental hazards by spreading the waste in thin layers, compacting these to the smallest practical volume, and applying and compacting a cover material of dirt at the end of each operating day.

sanitary sewer   Underground pipes that carry only domestic or commercial waste, not storm water.

sanitation   The control of those factors in the environment that can harmfully affect the development, health, or survival of humans.

sanitizer   Agent that reduces, but does not necessarily eliminate, all microorganisms that may be on a treated surface.

saprophyte   Parasite that lives on and derives nourishment from dead or decaying organic matter. A saprophytic organism is one that can obtain nourishment from nonliving organic materials.

SAR   Specific absorption rate.

SAR   Supplied-air respirator.

SARA Title III   Superfund Amendments and Reauthorization Act. Also known as the Emergency Planning and Community Right to Know Act of 1986.

sarcoma   A malignant tumor that develops from connective tissue cells.

sash   A moveable panel or door located at a laboratory hood inlet to form a protective shield and control the face velocity of air entering the hood.

saturated air   Air containing saturated water vapor with both the air and water vapor at the same dry-bulb temperature.

saturated hydrocarbon   Organic compound that is saturated with respect to hydrogen and cannot combine with atoms of other elements without giving up hydrogen.

saturated solution   A solution containing the maximum equilibrium amount of a solute at a specified temperature.

saturated steam   Steam at the boiling temperature corresponding to the pressure at which it exists.

saturation   The point at which the maximum amount of material can be held in solution at a given temperature.

Saybolt Universal seconds   [SSU or SUS]   Unit for measuring the viscosity of light petroleum products and lubricating oils. Also referred to as Saybolt seconds universal.

SBA   Small Business Administration.

SBS   Sick building syndrome.

s.c.   Subcutaneous.

scabies   Contagious skin disease caused by a mite and characterized by intense itching.

scald   To burn with hot liquid or steam.

scale   Deposits or an incrustation that may form on metals as a result of chemical or electrolytic action.

scanning electron microscope   [SEM]   A microscope that utilizes an electron beam that is directed at a sample to produce a reflected image of the sample material onto a screen from which fibers can be identified and counted.

scattered radiation   Ionizing radiation that, during its passage through a substance, has been deviated in direction and scattered by interaction with objects or within tissue. It may also have been modified by a decrease in energy.

SCBA   Self-contained breathing apparatus.

SCE   Sister chromatid exchange.

scf   Standard cubic foot (feet).

scf/d   Standard cubic feet per day.

scfm   Cubic feet of air per minute at standard conditions.

schistomiasis   An infection of humans resulting from the penetration of a worm through the skin, resulting in diarrhea, abdominal pain, urinary problems, liver damage, and possibly bladder cancer. The larvae are transmitted through impure water and infect both man and animals.

sciatica   Neuralgia (pain) of the sciatic nerve that is more typically referred to as a pain that radiates into either leg.

scientific notation   A form of expressing a number using a decimal value between 1 and 10 multiplied by a power of 10, such as 3.6×10^3, which is 3600.

scintillation counter   A device for determining the radioactivity of a material by interaction of radioactive emissions with a phosphor to produce light emissions, and a photomultiplier tube and electric circuits to facilitate counting the light emissions produced. A sodium iodide crystal is a common material used in scintillation counters. Also referred to as a scintillation detector.

sclera   The tough, white, outer layer of the eyeball.

scleroderma   Hardening of the skin.

sclerosis   The hardening of tissue with resulting loss of elasticity.

s/cm^3   Structures per cubic centimeter of air.

scotoma   An area of decreased vision within the visual field, surrounded by an area of normal vision.

SCP   Standards Completion Project.

scrubber   Pollution control device that uses a liquid to remove pollutants from emissions.

scuba   Self-contained underwater breathing apparatus.

sea breeze   A cool breeze blowing inland from the sea. Air movement toward the sea (i.e., off shore) in the evening and at night as a result of the water being warmer than the surrounding land. A local wind caused by uneven heating of land and ocean surfaces.

sealed source   A radioactive substance sealed in an impervious containment, such as a metal capsule, and that has sufficient mechanical strength to prevent direct contact with the radiation source or release of the radioactive substance from the containment, under normal conditions of use and wear. The NRC defines a sealed source as any byproduct material that is encased in a capsule designed to prevent leakage or escape of the byproduct material.

sebaceous glands   Glands which secrete sebum, a greasy lubricating substance.

seborrhea   An oily skin condition caused by an excess output of sebum from the sebaceous glands of the skin.

sebum   A thick, semifluid substance referred to as skin oil that is secreted by the sebaceous glands.

sec   Second.

SECALS   Separate Engineering Control Airborne Limits set by OSHA.

secondary air   Combustion air that is supplied over the firebed to oxidize the volatile constituents of the fuel, as distinguished from primary air that is introduced below the firebed.

secondary calibration method   Method that employs a device that must be calibrated against a primary standard method; not as accurate as a primary method. A wet-test meter, dry-gas meter, and a rotometer are examples of secondary standard methods that must be calibrated against a primary standard (*see* secondary standard).

secondary carcinogen   A carcinogen that needs metabolic activation; that is, one that is converted to a carcinogen by cell metabolism.

secondary combustion air   The air introduced above or below a fuel by natural, induced, or forced draft.

secondary infection   An infection acquired from person-to-person transfer from a primary case or from other secondary cases.

secondary pollutant   A pollutant formed in the atmosphere by chemical changes taking place between primary pollutants and other substances present in the air.

secondary radiation   Ionizing radiation originating as a result of the absorption of other radiation in matter.

secondary standard (Air Pollution)   An air pollution standard that establishes an ambient concentration of a pollutant with a margin of safety that will protect the environment from adverse effect.

secondary standard (Flow Rate)   Air flow measuring device that traces its calibration to a primary standard and that must be periodically recalibrated (*see* secondary calibration method).

secondary treatment (Wastewater)   Process and operation for converting dissolved organics to biological solids and removing those solids from the wastewater stream.

second-hand smoke   Tobacco smoke in the air that is inhaled by nonsmokers.

secular equilibrium (Ionizing Radiation)   The condition that exists when the ratio of parent nuclei to daughter nuclei remains constant with time. Thus, both parent and daughter decay at the same rate (i.e., that of the parent).

Sedgwick rafter cell   A glass slide/cell, formerly used to contain an aliquot of the collection media in which airborne particulate was collected. The cell was used to count the particulates microscopically so that a determination of dust concentration could be made.

sediment   Particulate matter that has been deposited in an area or has settled out of water or other liquid, e.g., urine.

sedimentation   The process by which solids settle out of a fluid (e.g., air or water).

sedimentation rate   Test that measures the rate at which blood cells (e.g., erythrocytes) settle out of a sample of drawn blood.

segregation (Public Health)   The separation for special consideration, control, or observation, of some part of a group of persons or of domestic animals from others, to facilitate the control of communicable diseases.

seismic data   Detailed information obtained from naturally or artificially produced earth vibration.

self-contained eyewash   An eyewash that is not permanently installed and must be refilled or replaced after use.

self-contained breathing apparatus   [SCBA]   *See* self-contained respirator.

self-contained respirator   One of three types of respiratory protective devices that is designed to provide breathing air to the wearer, independent of the surrounding atmosphere: open-circuit system, closed-circuit system with oxygen self-generating capability, and compressed air or oxygen closed-circuit device. Also classified as demand and pressure-demand units.

self-insured   The assumption of liability for worker compensation by a company to avoid cost of insurance provided by a private company.

SEM   Scanning electron microscope.

semantics   The study of the meaning or significance of words in a language.

semicircular canals   Special organs within the labyrinth of the inner ear that serve to maintain the sense of balance and orientation.

semi-conductor   Any of various solid crystalline substances, such as silicon, having electrical conductivity greater than insulators, but less than metals.

semi-conductor sensor   A sensor that responds to a contaminant that is present in the air as a result of its being adsorbed on the surface of the semiconductor-type sensor and producing a change in its conductivity in proportion to the concentration of the contaminant present in the sampled air.

semipermeable membrane   A barrier that permits the passage of some materials in a mixture, but not all.

senescence   The process of growing old.

senility   The state of being senile (old).

sensation   The awareness of the effects of a stimulus exciting one of the organs of the senses.

sense organ   A body organ that is sensitive to a specific stimulus, such as sight, hearing, smell, taste, and touch.

sensible   Capable of being perceived by one of the sense organs.

sensible heat   Heat that, when added or removed, results in a change of temperature.

sensitive   Capable of perceiving and responding to external stimuli or conditions.

sensitivity (Instrument)   The minimum amount of contaminant that can be repeatedly detected by the device, and the minimum input signal strength required to produce a desired value of output signal.

sensitivity (Physiology)   The ability of an organism to respond to stimuli.

sensitization   The process of rendering an individual sensitive to the action of a chemical. Involves an initial exposure of the individual to a specific antigen, resulting in an immune response. A subsequent exposure then induces a much stronger immune response.

sensitizer   A foreign agent or substance that is capable of causing a state of abnormal responsiveness in an individual. Following repeated or extended exposure to a substance, some people develop an allergic type of skin irritation referred to as sensitization dermatitis, while others may have a more serious response.

sensor (Instrument)   A transducer which converts a parameter at a sample point to a form suitable for measurement.

sensorineural hearing loss   Type of hearing loss typically caused by exposure to noise. This type hearing loss affects numerous people and is the result of damage to the inner ear, along with damage to the fibers of the acoustic nerve.

sensory response   The activity of the sense organs on stimulation that results in discomfort, but with no lasting or systemic injury.

SEP (OSHA)   Special Emphasis Program.

Separate Engineering Control Airborne Limits   [SECALS]   Separate exposure limits set by OSHA for industries in which it is not feasible to achieve the permissible exposure limit through engineering controls and work practices alone. Engineering and work practice controls must be used to reach the specified limit, and then the employer must provide respiratory protection to achieve the TWA-PEL.

sepsis   Infection of a wound or body tissues with bacteria, which leads to the formation of pus or to the multiplication of the bacteria in the blood.

septicemia   Blood poisoning, with actual growth of infectious organisms in the blood.

septic material   Material that can cause sepsis, which is the presence of pathogenic organisms or their toxins in the blood or tissues.

septic tank   An enclosure for the storage and processing of wastes where no sewer system exists. Bacteria decompose the organic matter into sludge, which must be pumped out periodically.

sequela   Condition, lesion, or any affection following or resulting from a disease. A pathological condition resulting from a disease.

serendipity   The accidental discovery of important new information.

serious bodily injury (OSHA)   An injury that involves a substantial risk of death, protracted unconsciousness, protracted and obvious physical disfigurement, or protracted loss or impairment of the function of a bodily member, organ, or mental facility.

serious hazard (OSHA)   Any condition or practice that could be classified as a serious violation of applicable federal or state statutes, regulations, or standards, based on criteria contained in the current Field Operations Manual or an approved state counterpart, except that the element of employer knowledge shall not be considered.

serious violation (OSHA)   Violation in which there is a substantial probability that death or serious physical harm could result from a condition that exists, and that the employer knew or should have known of the hazard.

serious violation citation (OSHA)   A citation that is issued when there is a substantial probability that death or serious physical harm could result from a condition that exists and the employer knows about or, with the exercise of reasonable diligence, could have known about.

serology   The branch of medicine concerned with the analysis of blood serum.

seropositive   The presence in the blood of antibodies to a disease-causing agent.

serpentine   One of the two major groups of minerals from which the asbestiform minerals are derived.

service life (Air-Purifying Respirator)   The period of time that a respirator, filter, or sorbent, or other respiratory equipment, provides adequate protection to the wearer.

settling chamber   A large chamber or expansion in an exhaust system in which the air velocity slows down to a rate at which larger particles suspended in the air settle out.

settling velocity   The velocity at which particles of specific sizes will settle out of the atmosphere due to the effect of gravity. Also referred to as the terminal velocity.

severity rate (Disabling Injury)   Relates the days charged to an accident with the hours worked during the period and expresses the result in terms of a million-hour unit.

sewage   Human body wastes and the wastes from toilets and other receptacles intended to receive or retain body wastes.

sewerage   The entire system of sewage collection, treatment, and disposal.

shakes   A worker term for metal fume fever.

shale oil   The hydrocarbon substance produced from the decomposition of kerogen when oil shale is heated in an oxygen-free environment. Raw shale oil resembles a heavy, viscous, low-sulfur crude oil.

sharps   Any sharp-pointed objects that have been in a health care setting and that, if the sharp end is contaminated by pathogens and pierces the skin, could transmit infection to a health care worker or other person.

Shaver's disease   A pneumoconiosis resulting from inhalation of fumes emitted from electric furnaces in the production of corundum.

sheaves   Grooved pulleys.

shield (Ionizing Radiation)   A material used to prevent or reduce the passage of ionizing radiation. Shielding is used to reduce the exposures of workers and patients to radiation emitted from sources.

SHI   Substance hazard index.

shift   The period of the day during which a person works.

shock   Any sudden emotional or physical disturbance. A physiologic response to bodily trauma, usually characterized by a rapid fall in blood pressure following an injury, operation, contact with electrical current, or other insult on the body.

short-circuiting   Situation that occurs when the air supplied for make up of air being exhausted flows to exhaust outlets (e.g., hoods, return air registers, etc.) before it enters or passes through the breathing/occupied zone.

short-term delayed effect   A health effect that occurs within a week to a month or so following a first exposure to a toxicant.

short-term exposure limit   [STEL]   A 15-minute, time-weighted average exposure to a substance that should not be exceeded at any time during a workday, even if the 8-hour exposure is within the TLV–TWA. Exposures up to the STEL value should not be for longer than 15 minutes, nor occur more than 4 times per day, and there should be at least 60 minutes between successive exposures in this range.

short ton   Equal to 2000 pounds or 907 kilograms.

shot blasting   A method for cleaning surfaces employing steel shot in a high-pressure air stream. A generic term for the cleaning of surfaces using abrasive cleaning agents.

SI   Systeme International, International System of Units.

siblings   Children borne by the same mother.

SIC   Standard Industrial Classification.

sick building syndrome   [SBS]   Situation in which building occupants experience acute health and/or discomfort effects that appear to be linked to time spent in the building, but in which no specific illness or cause can be identified.

sickness   Any condition or episode marked by pronounced deviation from the normal healthy state.

side shield   A device of metal or other material hinged or fixed firmly to a spectacle to protect the eye from side exposure.

side effect   An effect other than the intended one.

siderosis   A pneumoconiosis resulting from the inhalation of iron particulates.

sievert   [sv]   SI unit of any of the quantities expressed as dose equivalent. The dose equivalent in sieverts is equal to the absorbed dose in grays multiplied by the quality factor. One sievert is equal to 100 rems.

signal-to-noise ratio   The ratio of the desired signal to the undesired noise response measured in corresponding units.

silica   Crystalline silicon dioxide, occurring as quartz, tridymite, or cristobalite.

silica gel   A regenerative adsorbent material consisting of amorphous silica, which can be used as a solid sorbent for sampling some airborne contaminants.

silicosis   Pneumoconiosis due to the inhalation of dust containing silicon dioxide, resulting in the formation of generalized, nodular, fibrotic changes in the lung.

silicotuberculosis   A tuberculous infection of the silicotic lung.

silo-filler's disease   Pulmonary inflammation, often with acute pulmonary edema, resulting from the inhalation of irritant gases, particularly nitrogen dioxide, that collect in recently filled silos (i.e., from fresh green silage).

silt   Sediment carried or deposited by water.

silver solder   A brazing material that may contain cadmium. Exposure to cadmium fumes may occur if this filler metal contains cadmium.

simple asphyxiant   Physiological inert gases that, when present in the atmosphere in sufficient quantity, act to exclude an adequate supply of oxygen in the atmosphere being breathed.

simple sound source   A source that radiates sound uniformly in all directions under free-field conditions.

simulation   An image or representation of something, thus mimicking it.

single-use respirator   A disposable, maintenance-free respirator that is discarded after excessive resistance, sorbent exhaustion, or damage renders it unsuitable for further use.

sinus   Any of various air-filled cavities in the head, especially the one communicating with the nostrils.

SIP   State Implementation Plan.

SI Units   Systeme International d'Units.

skewed   A property of a statistical distribution indicating a lack of symmetry around the mean, such that the observations are concentrated to the left or right of the mean.

skewness   The tendency of a distribution to depart from symmetry around the mean.

skin carcinogen   A substance or physical agent that can produce skin cancer.

skin contamination   The presence of a hazardous substance on the skin, presenting the potential for irritation, corrosive action, sensitization, skin absorption, etc.

skin disorder   Any cutaneous abnormality or inflammation caused directly or indirectly by the work environment.

skin dose   The dose applied to the surface of the skin, or the dose received as a result of skin absorption.

skin injury   An immediate adverse effect on the skin that results from instantaneous trauma or brief exposure to toxic agents involving a single incident in the work environment.

skin notation   The reference on a TLV to the potential contribution to the overall exposure to the substance by absorption through the skin, mucous membranes, or the eye, upon contact with the material or its vapor.

slag   The fused and vitrified matter separated during the conversion of an ore to the metal product.

slag wool   Fibrous material made from the slag residue of the steel-making process. Similar to rock wool.

slimicide   A product used for the prevention or inhibition of the formation of biological slimes in industrial water cooling systems and for other applications.

sling   An assembly that connects the load to the material-handling equipment.

sling psychrometer   A device used to determine the properties of moist air by measuring the dry- and wet-bulb temperatures on thermometers fitted to a handle that enables their rapid rotation and the consequent evaporation of water from a wick placed over the bulb of the wet-bulb thermometer. The resulting temperatures (dry- and wet-bulb) are aligned on a psychrometric chart to determine the properties of the air.

slipped disc   A rupture or protrusion of a fibrocartilaginous inverterbral disc located between each of the vertebrae of the spine or vertebral column.

SLM   Sound level meter.

slot (Ventilation)   An opening in a hood for providing uniform air distribution.

slot velocity   Linear flow rate of air through an opening in a slot-type hood.

slough   An outer layer or covering that is shed, such as the dead outer skin.

sludge   Solid, semisolid, or liquid waste generated in wastewater treatment. Tarlike material that is formed when oil oxidizes. Oily residue with no commercial value. Wet residues removed from polluted water.

slurry   A mixture of liquid and finely divided insoluble materials.

slurry oil   The name applied to the heavy liquid stream obtained from the bottoms of the fluid catalytic cracking process employed in petroleum processing operations. This material has demonstrated carcinogenic effects to the skin in animal tests.

SMACNA   Sheet Metal and Air Conditioning Contractor's National Association.

s/cm³   Structures per cubic centimeter.

small quantity generator   A hazardous waste generator that produces less than 1000 kg of waste per month, or accumulates less than 1 kg of acutely hazardous waste at one time.

s/mm²   Structures per square millimeter.

smog   Irritating haze in the atmosphere as a result of the sun's effect on certain pollutants in the air, notably those from automobile and industrial exhaust emissions. Originally defined as a mixture of smoke and fog dispersed in the atmosphere.

smoke   An aerosol (air suspension) of particles, usually solids, that often originate or are formed by combustion or sublimation.

smoke detector   A device that senses visible or invisible particles of combustion.

smoke number   A dimensionless number that quantifies smoke emissions from a source.

smoke tube   A glass tube containing a chemical adsorbed on a solid media and that emits a smoke-like cloud when air is blown through the tube.

smolder   Burn with little smoke and no flame.

smooth muscle   Involuntary muscles lining the wall of the stomach, intestines, and arteries.

SMR   Standardized mortality ratio.

SMR   Standardized morbidity ratio.

soap bubble meter   *See* bubble meter.

soap film flowmeter   *See* bubble meter.

social drug   Alcoholic beverages and similarly used drugs.

Society of Toxicology   [SOT]   Professional association of toxicologists who have carried out original toxicity investigations, published findings, and who have a continuing professional interest in the field.

sociocusis   Increase in hearing-threshold level resulting from noise exposures in the social environment, exclusive of occupational-noise exposure, physiologic changes with age, or otologic disease.

SOCMA   Synthetic Organic Chemical Manufacturer's Association.

soft water   Water containing a low amount of minerals and other chemicals.

soil gas   Gaseous elements and compounds that occur in the small spaces between particles of the earth or soil.

soil gas sampling   A procedure used to locate oil and gas deposits and which has been adapted for use in the hazardous waste field in which volatile fuels or solvents are a concern. Passive and grab sampling techniques are employed. The former involves injecting a solvent into the soil to absorb the volatiles, and the latter involves the withdrawal of air and vapors through a probe. Analysis of samples is by gas chromatography or with a total gas analyzer.

solid sorbent   A solid-type sorbent material, such as activated charcoal, silica gel, porous polymer, etc., that is used to collect contaminants in air drawn through a tube containing the sorbent.

solid waste   Garbage, refuse, and sludge from waste treatment, solids from an air pollution control system, solids from waste water treatment facilities, discarded material, as well as wastes from industrial operations, farming, mining, and domestic activities.

solubility   A measure of the amount of a substance that will dissolve in a given amount of water or other material.

soluble   Capable of being dissolved.

solute   A substance dissolved in another substance, or the substance that is dissolved in a solvent.

solvent   Substance with a strong capability to dissolve another substance.

solvent extraction   A phase transfer process in which a substance is transferred from solution in one solvent to another without any chemical change taking place.

solvent–reagent blank response   The gross instrument response attributable to reagents and solvents used in preparing working standards for use in analytical procedures.

somatic   Pertaining to or characteristic of body tissue other than reproductive cells.

somatic cell   A body cell other than a reproductive cell usually with two sets of chromosomes.

somatic effect   Effects limited to the exposed individuals as distinguished from genetic effects to offspring.

somnolence   Unnatural drowsiness.

sone   A subjective unit of loudness equal to the loudness of a pure tone having a frequency of 1000 hertz at 40 decibels above the listener's hearing threshold.

sonic boom   The thunderlike noise reaching the ground when an airplane exceeds the speed of sound.

soot   Particulate formed from the incomplete combustion of carbonaceous matter and consisting of carbon combined with tar.

SOP   Standard operating procedure.

soporific   Substance that causes drowsiness or induces sleep.

sorbent   A general term for the solid or liquid materials that are employed to adsorb or absorb chemicals from air being passed through a bed or column of the material. Sorbents are used in respiratory protective equipment as well as in sampling devices (e.g., activated charcoal sampling tubes).

sorbent tube   A glass tube containing a sorbent material. Used in air monitoring to determine worker's exposure to vapors or gases.

sorption   The transfer of a substance from a solution to a solid phase, such as with the use of activated charcoal.

SOT   Society of Toxicology.

sound   Pressure variations that travel through the air and are detected by the ear. The auditory sensation evoked by an oscillation in pressure, stress, particle displacement, particle velocity, etc. in a medium with internal forces (e.g., elastic or viscous).

sound absorption   The change of sound energy into some other form of energy, such as heat, in passing through a material or striking a surface.

sound analyzer   A device for measuring the sound-pressure level as a function of frequency.

sound exposure   The cumulative acoustic stimulation at the ear of a person over a period of time.

sound field   A region containing sound waves.

sound intensity   The average rate at which sound energy is transmitted through a unit area, perpendicular to a specified point.

sound level   Weighted sound-pressure level determined with a sound level meter having a standard frequency-filter for attenuating part of the sound spectrum.

sound level contours   Lines drawn on a plot plan of a facility at positions of equal noise level.

sound level meter . An instrument comprised of a microphone, amplifier, frequency-weighting networks, and output meter that can be used to measure sound-pressure levels in a manner specified by the manufacturer.

sound power level (Acoustics)   [PWL]   Ten times the log to the base 10 of the ratio of a given power to a reference power (i.e., 1 E^{-12} watts).

sound pressure level (Acoustics)   [SPL]   The level, in decibels, of a sound equal to 20 times the logarithm to the base 10 of the ratio of the pressure of the sound to a reference pressure. The reference pressure is 2 E^{-4} microbar, which is equivalent to 20 micronewtons per meter squared.

sound shadow   The acoustical equivalent of a light shadow.

sound transmission loss   The ability of a barrier to block sound transmission as indicated by the reduction in decibels on the downstream side compared to that on the upstream side of the barrier material.

sound waves   The longitudinal waves moving out from a noise source and causing a gas, liquid, or solid to vibrate along the direction of the wave motion.

source (Ionizing Radiation)   A source of material that emits ionizing radiation as a result of radioactive decay (rather than electrically as in X-ray machines).

source material   Uranium or thorium, or any combination of these, in any physical form, or ores that contain by weight 0.05 percent or more of uranium, thorium, or any combination of these elements.

source of infection   The person, animal, object, or substance from which an infectious agent passes to a host.

source of injury   The object, substance, exposure, or bodily motion that directly produces or inflicts the injury.

sour crude oil   Crude oil that contains an appreciable quantity of hydrogen sulfide or other acid gas.

sour gas   A natural gas or other combustible gas that contains odor-causing sulfur compounds, such as hydrogen sulfide, mercaptans, etc.

SO$_x$ Sulfur oxides.

SP   Static pressure.

span (Instrument)   The algebraic difference between the upper and lower values of the range over which an instrument produces reliable results. Also expressed as the maximum value observable if the minimum is zero.

span drift (Instrument)   The change in the indicated response of an instrument over a specific time period of continuous operation due to causes other than a change in the concentration of the span gas. This drift can be positive or negative and may vary in magnitude between calibration periods.

span gas   A gas of known concentration that is used to calibrate or check the response of an instrument or analyzer.

spark   The rapid release of electrical or impact energy that is visible in the form of light.

spasm   A sudden, involuntary contraction of a muscle or group of muscles.

spasmodic   Relating to a sudden intermittent symptom.

spasmogenic   Causing spasms.

SPCC   Spill, prevention, containment, and countermeasures plan.

Special Emphasis Program   [SEP]   An OSHA program in which OSHA identifies a hazardous contaminant, such as silica, for priority rule-making, and then urges the using companies to establish programs to inform potentially exposed employees of the hazard and strategies for reducing exposures until rule-making can be promulgated.

special nuclear material   Material such as plutonium, uranium 233, uranium containing more than the natural abundance of uranium 235, or any material that has been enriched in any of the foregoing substances.

speciation (Chromatography)   Identifying and quantifying one or more components that are included in a chromatogram.

specific absorption rate   The absorption of radiofrequency or microwave radiation in watts per kilogram (W/kg) at specific frequencies.

specific activity (Ionizing Radiation)   Activity of a given radionuclide per unit mass (e.g., curies per gram) of the specific material.

specification   A clear and accurate description of the requirements for materials, products, services, etc., specifying the minimums of performance necessary for acceptability.

specific gravity   The ratio of the mass of a unit volume of a substance to the mass of the same volume of a standard substance at a standard temperature. Water at 39.2°F is the standard substance usually referred to.

specific heat   The amount of heat required to raise a unit weight of a substance 1 degree of temperature at constant pressure.

specific humidity   Weight of water vapor per pound of dry air.

specificity (Instrument)   The ability of an instrument to accurately detect a substance in the presence of others.

specific license (Ionizing Radiation)   A license that is issued by the NRC or an Agreement State to a company/organization to possess and use a radioactive material(s) after an application has been submitted and approved and specific requirements have been met.

specific volume   The volume occupied by a unit of air, such as cubic feet per pound.

spectra (Acoustics)   The distribution of noise energy according to frequency. Also referred to as an audio-frequency spectrum.

spectra (Electromagnetic Radiation)   The distribution of energy from a radiant source (e.g., visible light) according to its wavelength or frequency.

spectrograph   An analytical instrument used to photograph light spectra.

spectrophotometer   Instrument used to determine the distribution of energy within the spectrum of luminous radiation.

spectrophotometry   The selective absorption, by aqueous and other solutions, of definite wavelengths of light in the ultraviolet and visible regions of the electromagnetic spectrum as a means of determining the concentration of a substance present in the solution.

spectroscope   Instrument employed for observing, resolving, and recording the distribution of energy emitted by a source exposed to radiant energy.

spectrum (Acoustics)   A continuous range of sound components within which waves have some specified common characteristic, such as frequency, amplitude, or phase.

spectrum analyzer (Acoustics)   Instrument that indicates the frequency spectrum of an input signal.

specular surface   Resembling a mirrorlike surface.

speech interference level   [SIL]   The average, in decibels, of the sound pressure levels of a noise in the three octave bands of center frequency 500, 1000, and 2000 hertz. The result provides an indication of the effect of a noise on the reception of speech communication.

speed of sound   The speed at which sound travels, e.g., in air, it is 1178 feet per second at 78°F.

spent fuel (Nuclear Industry)   Reactor fuel that can no longer effectively sustain a nuclear reaction.

spent shale   Shale that remains after the kerogen present in oil shale has been converted to shale oil and removed.

sp gr   Specific gravity.

sphincter   A muscle that surrounds an orifice and functions to close it.

sphygmomanometer   A device for measuring arterial blood pressure.

spiked sample   A sample to which a known amount of substance has been added for the purpose of determining recovery or for quality control. Also called a spike.

spill   An unplanned release of a hazardous substance, such as a liquid, solid, gas, vapor, mist, or other form, that could result in worker's exposure to it or result in an adverse effect to the environment.

spill prevention, control, and countermeasures plan   [SPCC]   A plan that is developed and implemented to prevent spills of oil or hazardous substances from reaching navigable waters.

spirometer (Flow Calibration)   A primary standard for determining the flow rate of industrial hygiene sampling equipment and for calibrating secondary standards.

spirometer (Lung Function Test)   An instrument used to measure the volume of air taken into and exhaled from the lungs.

spirometry   The measurement of air or ventilatory capacity of the lungs.

SPL   Sound pressure level.

splash-proof goggles   Tight-fitting eye protection with indirect ventilation openings.

spoil   Dirt or rock that has been removed from its original location, destroying the composition of the soil in the process, as may occur in strip mining or dredging.

spontaneous combustion   The ignition of a material as a result of a heat-producing chemical reaction (exothermic) within the material itself, without exposure to an external source of ignition.

spontaneous ignition   Ignition resulting from a chemical reaction in which there is a slow generation of heat from the oxidation of a compound until the ignition temperature is reached. *See* also spontaneous combustion.

spontaneous ignition temperature   The temperature at which a material ignites of its own accord in the presence of air at standard conditions.

spore   A minute structure other than a seed that is capable of developing into a new growth. A microorganism, such as a bacterium, in a dormant or resting state.

sporicide   Substance that kills spores.

sprain   A joint injury in which some of the fibers of a supporting ligament are ruptured, but the continuity of the ligament remains intact.

spray booth   A semi-enclosure for automatic or manual spray application of protective coatings or paint to parts or fabricated products.

sputum   Material that is ejected from the lungs, bronchi, and trachea through the mouth.

sputum cytology   The investigation of sputum samples as a screening test for lung cancer in high-risk populations.

sq. ft.   Square foot (feet).

SQG   Small quantity generator (of hazardous waste).

squamous   Covered with or formed of scales.

squamous cells   Flat or scalelike epithelial cells.

squamous cell carcinoma   A skin tumor that appears as a nodule or a red scaly patch on the ears, face, lips, or mouth.

square foot   [sq. ft.]   A unit of area equal to 144 square inches or 0.0929 square meters.

square meter   [sq. m.]   A unit of area equal to 10,000 square centimeters or 10.76 square feet.

squirrel cage fan   A centrifugal blower with forward-curved blades.

SSOV   Safety shutoff valve.

SSU   Saybolt Universal seconds (Also abbreviated as SUS.)

stability (Atmospheric)   The tendency of the atmosphere to resist vertical motion, or alternately, to supress existing turbulence. It is related to both wind shear and temperature structure vertically, but the latter is generally used as an indicator of atmospheric stability.

stable air   An air mass that stays in the same position rather than moving in its normal direction, thereby resulting in a buildup of air pollutants.

stable material   Material that normally has the capacity to resist changes in its chemical composition, despite exposure to air, water, and heat as encountered in fire emergencies.

stack   The device at the end of a ventilation system or furnace through which exhaust from the operation or ventilation system is released to the atmosphere.

stack effect   Pressure-driven airflow produced by convection as heated air rises, creating a positive pressure in the top of a building and a negative pressure at the bottom. In houses and buildings it is the tendency toward the displacement of internal heated air by unheated outside air due to the difference in density of the outside and inside air.

stack sampling   The collection of representative samples of gaseous or particulate matter flowing through a stack or duct.

stagnation (Air Pollution)   An atmospheric condition in which there is a lack of air movement, resulting in a buildup of air contaminants.

stamen   The part of a flower or plant that produces pollen.

standard   Something that serves as a basis for comparison.

standard air (Industrial Hygiene)   Air at 25°C (77°F) and 760 millimeters of mercury pressure.

standard air (Ventilation)   Air at 70°F, 50 percent relative humidity, 29.92 inches of mercury atmospheric pressure, and weighing 0.075 pounds per cubic foot.

standard air density (Ventilation)   The density of air at standard conditions is 0.075 pounds per cubic foot.

standard cubic foot   A volume unit of measurement at a specified temperature and pressure. The temperature and/or pressure vary based on the discipline. For example, the specified temperature employed in industrial hygiene determinations is 25°C.

standard deviation (Sample)   A unitless number that indicates the scatter of data from the mean. A measure of the variability or dispersion of a set of results. The square root of the sample variance.

standard error   An estimate of the magnitude by which an obtained value differs from the true value.

Standard Industrial Classification   [SIC]   Code for classifying all types of commercial businesses (including industry) based on their primary product or service rendered.

standardized mortality ratio   [SMR]   The ratio of the number of deaths observed in a study group (cohort) to the number of deaths expected in the study group based on the rate as determined for an unexposed control population. If the ratio of these results is greater than 1, it indicates an increased risk for the exposed population. The greater the SMR is above 1, the greater the risk.

standard man   A theoretical, physically fit man of standard height, weight, and other parameters, including blood, tissue composition, percent water, weight of organs, etc., that can be used in studies of man's response to various stimuli and for designing equipment relative to ergonomic considerations. Also referred to as reference man.

standard operating procedure (Process Industry)   Documents setting forth the operating procedures covering all details currently in effect for an operation.

Standards Completion Project   A NIOSH-supported project that was carried out to develop sampling and analytical methods for application in the field of industrial hygiene.

standard score   The number of standard deviations that a given value is above or below the mean. Also called the Z score.

standard solution   A solution of known concentration that can be used in the preparation or measurement of other solutions or chemicals.

standard temperature and pressure (Industrial Hygiene)   A temperature of 25°C and a pressure of 760 millimeters of mercury.

standard temperature and pressure (Physics)   [STP]   0°C and 1 atmosphere.

standard threshold shift (OSHA)   A change in hearing threshold relative to the baseline audiogram of an average of 10 decibels or more at the 2000, 3000, and 4000 hertz frequencies in either ear.

standing wave (Acoustics)   A periodic wave having a fixed distribution in space that is the result of interference of progressive waves of the same frequency and kind. In such situations sound does not decrease as the distance from the source is increased. Marked variations in sound pressure are observed. The measured sound pressure decreases to a minimum, rises again to a maximum, decreases to a minimum, increases to a maximum, etc. Such patterns are referred to as standing waves.

standpipe   A vertical pipe into which water is pumped in order to produce a desired pressure.

stannosis   A form of a pneumoconiosis resulting from the inhalation of tin-bearing dust.

stasis   The stoppage or lessening of the flow of blood or other body fluid in any part of the body.

State Implementation Plan   [SIP]   The method or program that a state will implement and follow to meet an EPA air standard.

static electricity   Literally, electricity at rest; consists of opposite electrical charges that are usually kept apart by insulation; the result of the accumulation of electric charge on an insulated body and its potential for discharge as a result of such an accumulation of electric charge.

static pressure   The potential pressure exerted in all directions by a fluid at rest. When above atmospheric pressure, it is positive; when below atmospheric pressure, it is negative.

stationary source   Any facility, building, or installation which emits or may emit an air pollutant that is subject to regulation under the Clean Air Act.

statistic   A characteristic of a population or a sample of it, such as the mean, variance, etc.

statistical inference   The process of predicting population variables based on sample data.

statistical significance   An inference that the probability is low that the observed difference in quantities being evaluated could be due to variability in the data rather than an actual difference in the quantities. The inference that an observed difference is statistically significant is typically based on a test to reject one hypothesis and accept another.

statistical test   A procedure that enables one to decide whether an hypothesis about a distribution of one or more populations or variables should be rejected or accepted.

statistics   The collection, organization, description, and analysis of data.

std   Standard.

STD   Sexually transmitted disease.

steady state   The condition of a system when the inflow of materials or energy equals the output.

steady-state noise   Sounds that remain constant with time, such as that of an air conditioner when in operation.

steam   Water in its gaseous state.

steam-fitter's asthma   Asthma resulting from exposure to asbestos. It is accompanied by asbestosis.

STEL   Short-term exposure limit.

stenosis   Narrowing of a body passage or opening.

sterile   Free from living organisms.

sterilization   The use of a chemical or other means for the complete elimination of microbial life.

sterilize   To make free of microorganisms.

stethoscope   A device for listening to sounds in the chest.

still air   Air with a velocity of 25 feet per minute or less.

sting   A sharp momentary pain, typically the result of a puncture of the skin by an insect or other arthropod.

stink damp   Hydrogen sulfide.

stipple cell   A red blood cell containing granules of varying size and shape.

stochastic effects   Health effects that occur randomly and for which the probability of the effect occurring, rather than its severity, is assumed to be a linear function of dose without threshold. Hereditary effects and cancer incidence are examples of stochastic effects.

stock solution   Solution consisting of an accurately measured weight of a substance of known purity dissolved in a known volume of suitable solvent. The concentration of this solution is traceable to a primary weight standard.

Stoddard solvent   Solvent with a minimum flash point of 100°F (37.8°C), relatively low odor level, and other properties that conform to an ASTM test method.

stoichiometric   The exact proportion of two or more substances that will permit a chemical reaction with none of the individual reactants left over.

stoichiometry   A branch of chemistry dealing with reactions in terms of the relative quantities of each reactant and each product.

Stoke's law   Regarding the fall of a liquid or solid body through any fluid media, this law states that the settling velocity is a function of gravity, the diameter and density of the falling body, and the coefficient of viscosity, viscosity, and density of the media.

stomatitis   Inflammation of the oral mucosa (mucous lining), due to local or systemic factors.

stone mason's disease   Silicosis.

storage loss   An estimate of the typical losses that could occur due to the storage of samples for prolonged time periods (e.g., 10, 15, 20, etc. days) as determined using samples obtained from the same test atmosphere. *See* also storage sample stability.

story   That portion of a building structure included between the upper surface of a floor and the upper surface of the floor or roof above.

STP   Standard temperature and pressure.

strain   To effect a change as a result of the application of a stress. Physiologic, psychologic, or behavioral manifestation of stress on the body.

strategy   A plan of action to accomplish a stated goal.

stratum corneum   External layer of the epidermis.

streamline flow   A flow that exists when fluid (e.g., air) and particles are moving in a straight line parallel to the axis of a pipe or duct.

streamlines (Ventilation)   Lines perpendicular to velocity contours.

stress   A physical, chemical, or emotional factor that causes bodily or mental tension and may be a factor in disease causation, fatigue, or strain. The response of the body to a demand made on it.

stressor   Agent, condition, or thing that causes stress on the body.

stroboscopic effect (Illumination)   The effect noticed when rapidly moving objects, when observed under fluorescent or mercury lighting systems, appear to be blurred or not moving.

struck-against injury   Injury in which a worker collides with a stationary object.

struck-by injury   Injury in which a worker is hit by a moving object or particle.

structure-borne sound   Sound that travels over at least a part of its path by means of the vibration of a solid structure.

structures (Asbestos)   A microscopic bundle, cluster, or matrix made up of asbestos fibers or that may contain asbestos fibers.

structures per cubic centimeter of air   [s/cm^3]   The number of asbestos structures determined to be present in one cubic centimeter of air. Typically, there are more asbestos structures per cubic centimeter of air seen on a filter sample than there are fibers per cc of air because the analytical method of counting structures (TEM) sees more countable shapes than are seen by the phase contrast method of counting fibers.

stupor   Partial or near-complete unconsciousness.

styrene sickness   Nausea, vomiting, and fatigue resulting from exposure to styrene.

subacute   An effect between acute and chronic.

subatomic particles   A group of particles that includes positively charged protons, negatively charged electrons, and uncharged neutrons.

subclinical   The early stages of a disease without clinical signs.

subcutaneous   Beneath the skin.

subjective   Affected by personal bias or opinion.

sublimation   The process of passing from a solid state directly to a gaseous state.

subliminal   Too weak to cause a sensation; that is, below the threshold of perception.

sublingual   Below the tongue.

subsidence   A settling or sinking of the surface of the earth due to natural or man-made factors.

subsonic sound   Sound energy in the frequency range below 20 hertz.

substance hazard index   [SHI]   The ratio of the vapor pressure of a substance to its acute toxicity.

substernal   Beneath the breastbone.

substitution   The replacement of a hazardous material or source of physical stress with a less hazardous one.

substrate   The material underneath or to which another is attached.

subsystem   A part of a system, such as a motor in a piece of machinery, or a segment of a production process.

sudorific   Causing an increase in sweating.

suffocate   To impair breathing or respiration.

SUMMA canister   An evacuated stainless steel canister used for collecting samples of ambient air.

sump   A depression or tank that catches and retains liquid runoff for subsequent disposal.

Superfund   A federal program which taxes chemical production to raise money which is then used to clean up toxic waste sites.

Superfund Amendments and Re-Authorization Act   A law enacted to ensure that communities throughout the country would be prepared to respond to chemical accidents and to provide the public with information on hazardous and toxic chemicals used and released in their communities (SARA Title III). Also known as the Emergency Planning and Community Right to Know Act of 1986.

superheated steam   Steam at a temperature higher than the boiling temperature corresponding to the pressure at which it exists.

supination   The rotation of the forearm about its longitudinal axis.

supplied air respirator   A respirator that has a central source of breathing air supplied to the wearer via an airline.

supply air   Air supplied to a space by an air handling system to be used for ventilating, heating, cooling, humidification, or dehumidification.

supply air diffuser   A fixture and opening through which air is supplied to a space.

suppuration   The formation of pus, or the act of becoming converted into and discharging pus.

surface tension   The attractive force exerted by the molecules below the surface of a liquid upon those at the surface/air interface, as a result of which the surface tends to contract and has properties resembling those of a stretched elastic membrane.

surface water   Any of the water that is on the surface of the earth.

surfacing ACM   An asbestos-containing material that is sprayed on, trowelled on, or otherwise applied to surfaces such as ceilings or structural members for fireproofing or acoustical purposes.

surfactant   A surface-active agent that can be added to water to improve the ability of the water to wet a material or surface.

surgical waste   Waste generated by surgical procedures in the treatment of diseases, injuries, or other medical procedures.

surrogate (Sampling)   The measurement of one compound in place of determining the presence of many others. For example, the determination of an oxygenate component of gasoline as a surrogate or marker for that product rather than determining the presence of all components in it.

surveillance (Public Health)   The practice of close supervision of contacts for purposes of prompt recognition of infection or illness, but without restricting their movement.

survey (Industrial Hygiene)   The determination of the exposure of workers to health hazards based on the concentration, frequency, and duration of exposure, as well as the exposure controls and work practices associated with an individual's exposure to an airborne contaminant, physical stress, ergonomic factor, or biologic agent.

survey (Ionizing Radiation)   An evaluation of the radiological conditions and potential hazards associated with the production, use, transfer, release, disposal, or presence of a radioactive material or other source of exposure to ionizing radiation.

survey meter   A portable instrument that measures ionizing radiation dose rate and provides an estimate of the amount of radioactive material present.

susceptible (General)   Description of a person or animal presumably not possessing resistance against a particular pathogenic agent and who, for that reason, is liable to contract the disease if exposed to the agent.

suspect carcinogen   A material that is believed to be capable of causing cancer, but for which there is limited scientific evidence.

suspended solids   Small particles that are dispersed but not dissolved in a liquid or gas.

Sv   Sievert.

SVOC   Semi-volatile organic compound.

sweat glands   Glands in the skin that produce sweat.

sweating   The excretion of perspiration through the pores of the skin.

sweep guard   A single or double arm that is attached to the upper die or slide of a press for moving the operator's hands to a safe position as the dies close.

sweet crude oil   Crude oil that is low in sulfur, especially containing little or no hydrogen sulfide.

swimmer's itch   A dermatitis due to the penetration of the skin by the larval form of schistosomes.

symbiosis   The relationship of two or more organisms in an association that may be mutually beneficial to each.

symptom   Subjective evidence of disease or of an individual's condition as perceived by the person.

symptomatic   Pertaining to or indicative of a disease or disorder.

syncope   A temporary suspension of consciousness; fainting.

syndrome   A group of signs and symptoms that collectively indicate or characterize a disease or abnormal condition. A set of symptoms that occur together.

synergism   Cooperative action of two or more substances whose effect is greater than the sum of their separate effects.

synergistic   Acting together.

synergistic effect   The added effect produced by two processes working together in combination, the result of which is greater than the sum of the individual effects.

synergy   The effect when two or more substances, conditions, organisms, etc. achieve a result that each is incapable of achieving individually.

synonym   Another name by which a specific chemical may be known. For example, a synonym for toluene is toluol.

synovitis   Inflammation of a synovial membrane (i.e., secretes synovia).

synthesis   The reaction or series of reactions by which a complex compound is obtained from simpler compounds or elements.

system   A set or arrangement of components so related or connected as to form a unity or whole.

systematic errors   Errors introduced by an individual, the result of a poor method/technique, improper reading/recording of data, or a consistent error in an instrument. These do not cancel out if more samples are collected and analyzed, and they always cause bias.

Systeme International d'Units   [SI]   Metric-based system of weights and measures adopted by many countries, including the U.S.

systemic   Affecting the body and/or the organs that are remote from the site of exposure.

systemic effect   A toxic effect that is remote from the point of contact or site at which the material entered the body. For example, vinyl chloride enters the body by inhalation, but affects the liver if a sufficient dose is absorbed by this route.

systemic toxicity   Adverse effects that are caused by a substance that affects the body in general rather than in a local manner. For example, the effect of a chemical substance in an area remote from the site of entry into the body.

system safety   A program oriented toward hardware and the functioning of systems, subsystems, and the components of the system and subsystems.

systole   Heart muscle contraction, especially that of the ventricles.

T

t   Temperature.

t   Tonne.

T   Tera, 1 E^{12}.

T   Tesla.

T   Ton.

t$_{1/2}$ Physical or radiological half-life.

tachometer   A device for determining rotational speed.

tachycardia   Excessive rapidity in the action of the heart, i.e., a rapid heart rate.

tachypnea   Increased rate of respiration.

tactile   Pertaining to the sense of touch.

tagout device   A prominent warning device that is capable of being securely attached to equipment start-up devices and that, for the purpose of protecting personnel, forbids the operation of an energy-isolating device and identifies the applier or authority who has control of the warning device.

tagout procedure   The practice of affixing a warning tag to indicate that a piece of equipment or device should not be started and/or used.

tailings   The wastes separated from mineral ores during processing.

talc   A hydrated magnesium silicate material similar in chemical composition to asbestos. It is generally a flaky mineral material, but may also be fibrous. Some talcs contain asbestos in small amounts.

talcosis   A pneumoconiosis resulting from the prolonged inhalation of talc dust.

tape sampler   A type of air sampling device that draws air through a chemically treated filter paper or a fine filter paper, and is used for the measurement of gases and fine particles, allowing air sampling to occur over a set period of time automatically at predetermined times.

tare weight   The weight of a container, liner, wrapper, or sampling media (e.g., a filter) before sampling through it, which is then deducted from the final weight to determine the net weight of a collected material (e.g., particulates collected on a filter during sampling).

target dose   The amount of a substance that reaches the site of action, i.e. organ.

target organ   The organ of the body that is most affected by exposure to a particular substance.

target organ effect   The health effect on a specific organ as a result of an exposure to a specific substance.

target organ toxin   A toxic substance that attacks a specific organ of the body.

tar sand   Sandstone that contains very heavy, tarlike hydrocarbons.

task illumination   The minimum amount of illumination necessary to carry out a task.

taxonomy   The theory, principles, and process of classifying organisms in established categories.

t_b Biological half-life.

T.B.   Tuberculosis.

TBS   Tight building syndrome.

TC   Thermal conductivity.

TC   Toxicity characteristic.

TCAP   Toxic Catastrophic Prevention Act Program.

TCD   Thermal conductivity detector.

Tcf   Trillion cubic feet, 1 E^{12} ft^3.

TCLP   Toxicity characteristic leaching procedure.

TECP suit   Totally-encapsulated chemical protective suit.

telemetry   The transmission of data collected at a location over communication channels to a central station.

TEM   Transmission electron microscope.

temperature   A measure of the intensity of heat expressed in degrees Celsius or Fahrenheit.

temperature effect (Sampling)   The effect of air temperature on the response of an instrument to a contaminant/stress factor being measured, or the effect of temperature on the adsorption/absorption of a contaminant by a collecting media. Typically, effects are greatest at temperature extremes, such as below 0°C and above about 40°C.

temperature gradient   The rate of change of temperature with displacement in a given direction.

temperature inversion   Vertical temperature distribution such that temperature increases with height above the ground.

tempered air   Supply air that has been heated sufficiently to prevent cold drafts.

template   A pattern or guide for making or doing something.

temporary partial disability (Worker's Compensation)   A condition in which there is a partial loss of earning power, but from which full recovery is expected.

temporary threshold shift (Acoustics)   [TTS]   A temporary impairment of hearing ability as indicated by an increase in the threshold of audibility.

temporary total disability (Worker's Compensation)   A condition under which an employee is unable to work, but from which a complete or partial recovery is expected.

tendinitis   Inflammation of tendons and of tendon–muscle attachments.

tennis elbow   *See* epicondylitis.

tenosynovitis   Inflammation of the connective tissue sheath of a tendon.

tenth-value layer (Ionizing Radiation)   The thickness of an absorber needed to reduce a radiation field to one-tenth its incident value.

tepid   Lukewarm water at a temperature between 80 and 95°F.

tera   [T]   Prefix designating 1 E^{12}.

teratogen   An agent or factor that causes the production of physical defects in a developing embryo. A stressor, the exposure to which may potentially result in fetal health effects.

teratogenesis   The process whereby abnormalities of the offspring are generated, usually as the result of damage to the embryonal structure during the first trimester of pregnancy, producing deformity of the fetus.

teratogenic   Able to or tending to produce malformations.

teratogenicity   The ability of a substance to produce a teratogenic effect.

teratology   The study of malformations or deviations from that which is morphologically and functionally normal in developing and growing organisms.

terminal settling velocity   *See* terminal velocity.

terminal velocity   The velocity of a falling particle when the frictional drag equals the gravitational force, and the particle falls at a constant rate. *See* settling velocity.

tertiary treatment (Wastewater)   Treatment that involves the removal of suspended solids and/or nutrients, typically compounds of phosphorous or nitrogen, from wastewater.

tesla   [T]   The unit of magnetic flux density in the International System equal to one weber per square meter.

test gas   A gaseous contaminant that has been diluted with clean air (or nitrogen in some cases) to a known concentration.

tetanus   An acute, often fatal infection, caused by a bacillus that generally enters the body through a wound. The disease is characterized by rigidity and spasmodic contractions of voluntary muscles. See tetany.

tetany   A condition in humans marked by severe, intermittent, involuntary, muscular contractions and pain.

THC (Gas Chromatography)   Total hydrocarbons.

theoretical air   The quantity of air, calculated from the chemical composition of the material to be combusted, that is required to burn it completely.

theory   An accepted principle or explanation of a phenomenon that is supported by evidence.

therapeutics   The science and art of treating disease.

therapy   The treatment of illness or disability.

therm   A measure of heat content equal to 100,000 Btu.

thermal anemometer   A device used to measure fluid velocity (i.e., air) by sensing the changes to heat transfer from a small electrically heated sensor exposed to the fluid flow.

thermal comfort (Indoor Air Quality)   The combination of air temperature, air velocity, mean radiant temperature, and air humidity that is comfortable to 90 percent of the occupants. Some standards indicate thermal comfort exists when 80 percent of occupants are satisfied with conditions.

thermal conductivity detector   [TC detector]   A detector that measures the specific heat of conductance as a quantitative means to determine the concentration of a substance.

thermal decomposition   A chemical breakdown of a material as a result of exposure to heat. The decomposition products are often more toxic than was the parent material.

thermal pollution   A change in the quality of an environment as a result of raising its temperature. The discharge of a source of heat to a body of water such that the increase in the temperature of the receiving body depletes oxygen and adversely affects the environment of aquatic organisms, fish, etc.

thermal radiation   The transfer of heat by means of electromagnetic waves.

thermal system insulation   [TSI]   Insulation used on pipes, fittings, boilers, ducts, etc. to prevent heat loss or water condensation.

thermistor   A semi-conductor that exhibits rapid and large changes in resistance for relatively small changes in temperature and that is used to measure temperature.

thermoanemometer   Device for measuring air velocity. Also referred to as a heated wire anemometer or heated thermocouple anemometer.

thermocouple   A thermoelectric device, consisting of two dissimilar metals, that can be used to measure temperature, or the effect of a change in temperature, as a result of a difference in electrical potential between the metals when exposed to heat.

thermodynamics   The science concerned with heat and work and the relationship between them.

thermograph   A temperature measuring system that gives a graphic record of the time variation of temperature.

thermolabile   Changed or destroyed by high temperature.

thermoluminescent dosimeter   A radiation badge worn by a person to measure radiation exposure dose. It contains a radiation-sensitive crystalline material that emits light on exposure to a heat source in proportion to the amount of radiation absorbed.

thermophilic bacteria   Organisms that thrive at temperatures exceeding 55°C.

thermoplastic   A plastic material that is capable of being repeatedly softened by heat and hardened when cooled.

thermosetting plastic   Plastic that hardens when first heated under pressure, but whose original characteristics are destroyed when remelted or remolded.

thermostable   Substance that is resistant to changes by heating.

thermostat   A control device for regulating temperature.

thief   A device that is lowered into a tank to take a sample of the stored material from any desired depth.

third-octave band analysis (Acoustics)   A spectral analysis of noise in which the data are plotted as the rms level in a band of frequencies one-third octave wide as a function of the center frequency of the band.

thixotropy   The property exhibited by a fluid that is in a liquid state when flowing and in a semisolid gel state when at rest.

thoracic cavity   The space above the diaphragm and below the neck.

thoracic fraction   The mass fraction of inhaled particles that penetrate beyond the larynx.

Thoracic Particulate Mass   Those particles that penetrate a separator whose size collection efficiency is described by a cumulative lognormal function with a median aerodynamic diameter of 10 micrometers and with a geometric standard deviation of 1.5.

Thoracic Particulate Mass TLVs [TPM TLVs] Exposure limits assigned to those materials that are hazardous when deposited anywhere within the lung airways and the gas–exchange region of the lungs.

thorium series Isotopes that belong to a chain of successive decays that begin with thorium 232 and ends with lead 206.

three E's of safety Engineering hazards out of the job; Education of workers; and Enforcement of safety rules/practices.

threshold The intensity of a stimulus that is just sufficient to produce a sensation or elicit a response. The level at which effects are observed.

threshold audiometry Procedure used for determining a worker's hearing threshold to determine hearing acuity and the presence of an abnormal function of the ear.

threshold dose The minimum amount of a substance required to produce a measurable effect.

threshold hypothesis (Ionizing Radiation) An assumption that no injury occurs below a specified dose of radiation.

threshold limit value [TLV] The airborne concentration of a substance and the conditions under which it is believed that nearly all workers may be repeatedly exposed day after day without adverse health effects. Because of wide variation in individual susceptibility, however, a small percentage of workers may experience discomfort from some substances at concentrations at or below the threshold limit; a smaller percentage may be affected more seriously by aggravation of preexisting conditions or by development of occupational illnesses. A copyrighted term of the American Conference of Governmental Industrial Hygienists.

threshold limit value–ceiling [TLV–C] The concentration that should not be exceeded during any part of the working exposure.

threshold limit value–short-term exposure limit [TLV–STEL] The concentration of a substance to which workers can be exposed continuously for a short period of time without suffering irritation, chronic or irreversible tissue damage, or narcosis of sufficient degree to increase the likelihood of accidental injury, impair self-rescue, or materially reduce work efficiency; and provided the TLV–TWA is not exceeded.

threshold limit value–time-weighted average [TLV–TWA] The values of toxic materials in air that are to be used as guides for the control of health hazards in the work environment. They represent time-weighted average concentrations to which nearly all workers can be exposed for a normal 8-hour workday and 40-hour workweek, day after day, without adverse health effects.

threshold of audibility Minimum sound-pressure level capable of evoking an auditory sensation in a specified number of trials. *See* hearing threshold.

threshold of pain (Acoustics) The level of sounds sufficiently high to cause pain (e.g., about 140 decibels and above).

threshold of pain (Heat) The level of heat on surfaces so that contact with them is enough to cause pain; or the level of radiant heat sources that are intense enough to result in skin/surface temperatures that cause pain. Temperatures of 125°F and above cause pain on contact.

Threshold Planning Quantity [TPQ] The amount of material at a facility that requires emergency planning and notification per CERCLA.

threshold shift (Acoustics)   An increase in the hearing threshold level that results from exposure to noise.

threshold toxicant   A substance that is known or assumed to produce no adverse effect below a certain dose or dose rate.

throb   A pulsating, rhythmic sensation or movement.

thrombocytopenia   A decrease in the number of platelets in the blood.

throw (Ventilation)   The distance from an air supply opening, measured in the direction of flow, to the point where the air velocity is 50 feet per minute.

tic   A muscle spasm of brief duration.

tick   Bloodsucking parasitic insect that can transmit infectious diseases.

tidal volume   [TV]   The volume of air that is inspired or expired in a single breath during breathing.

tight building syndrome   [TBS]   A condition associated with buildings designed and operated at minimum outdoor air supply that, as a result, often leads to complaints of adverse health effects and/or discomfort by the occupants. Also referred to as SBS (sick building syndrome).

tight-fitting facepiece   A respiratory inlet covering that forms a complete seal with the face.

TIG welding   Tungsten inert-gas welding.

time-weighted average exposure   [TWA]   A worker's average exposure to an air contaminant, physical stress, or bioorganism over the work period or shift. Also referred to as the time-weighted average concentration.

time-weighted average sound level   That sound level which, if constant over an 8-hour exposure, would result in the same noise dose as measured.

tingling   A stinging sensation.

tinnitus   Head noises and noise in the ears, such as ringing, buzzing, roaring, or clicking. Such sounds may be heard, at times, by other than the patient.

titration   The determination of a constituent in a known volume of a solution by the measured addition of a second solution of known strength to the completion of its reaction with the component in the first solution, as indicated, typically, by the formation of a colored end point.

TLC   Thin-layer chromatography.

TLD   Thermoluminescent dosimeter.

TLV   Threshold limit value.

TLV–C   Threshold limit value–ceiling.

TLV–STEL   Threshold limit value–short-term exposure limit.

TLV–TWA   threshold limit value–time-weighted average.

toeboard   A vertical barrier at floor level that is erected along exposed edges of a floor opening, platform, runway, or ramp to prevent falls of materials.

tolerance   The ability to endure an unusual amount of stress or dose of a substance that would typically adversely affect others.

tolerance limit   The greatest amount or intensity of a biological agent, chemical, or stimulus that a living system can withstand for a specified time period and not show evidence of injury or damage and still survive.

ton   A unit of weight in the U.S. Customary System equal to 2000 pounds. Also referred to as a short ton.

tone (Acoustics)   A sound sensation having a pitch.

tone (Physiology)   The continuous partial contraction of a muscle.

tone deaf   The inability to discriminate between tones that are close together in pitch.

tonne   A mass in the metric system equal to 1000 kilograms. Also referred to as a long ton.

ton of refrigeration   The extraction of 12,000 Btu per hour or 288,000 Btu per day of 24 hours. The latter is referred to as a ton-day of refrigeration. Also referred to as a ton of air conditioning.

topography   The physical characteristics of a surface area, including relative elevations and the position of natural and man-made features.

topping off   The operation of completing the loading of a tank to a required ullage.

torr   A unit of pressure equal to 1.316 E^{-3} atmospheres. A pressure of 760 torr is equal to 1 atmosphere of pressure; thus, a torr is equal to 1 millimeter of mercury.

total airborne particulate   *See* total particulate.

total body burden   The amount of a substance in the body as a result of previous exposure(s) to the substance.

total dissolved solids (Wastewater)   The filterable solids in the wastewater.

total hydrocarbons (Gas Chromatography)   The total area or counts of an integrator in response to the output signal of a gas chromatograph detector (e.g., flame ionization detector) when analyzing a sample containing a mixture of hydrocarbons, such as is present in gasoline. Typically, the count is equated to that from a known amount of a hydrocarbon, such as n-hexane, to report the total hydrocarbon concentration.

total lung capacity   [TLC]   The volume of air in the lungs following maximum inspiration.

totally disabled   A worker who is injured and cannot do any work because of the injury.

totally-encapsulated chemical protective suit   A full-body garment constructed of chemical protective materials that covers the wearer's torso, head, arms, legs, and respirator, and may cover the hands and feet with tightly attached gloves and boots. It completely encloses the wearer and respirator by itself or in combination with the wearer's gloves and boots.

total oxygen demand   The combination of the biological and chemical oxygen demand of a body of water, or of sewage or an organic waste stream.

total particulate   The concentration of particulates in air without respect to the size of the particles collected. The concentration is expressed in milligrams per cubic meter of air. Also referred to as total airborne particulate.

total pressure   [TP]   The algebraic sum of static pressure and velocity pressure of a fluid with regard for sign.

total residual chlorine   [TRC]   The amount of chlorine remaining in water following treatment with chlorine or a chlorine compound, the purpose of which is to inhibit the regrowth of bacteria in treated water.

total solids   The total amount of dissolved and undissolved constituents in water or wastewater.

total suspended particulate   [TSP]   A weight determination of the particulate matter in the ambient air as determined from a filter sample obtained using a high volume air sampler.

toxemia   A condition in which toxins produced by body cells at a local site of infection are contained in the blood. This condition is also referred to as blood poisoning.

toxic   Pertaining to, due to, or of the nature of a poison. Having poisonous effects.

toxicant   A poisonous agent.

toxic dose   The dose required to produce a toxic effect.

toxic dust   Dust that may be harmful to the respiratory system or other part of the body if inhaled and/or absorbed into the bloodstream.

toxic effect   The effect produced by a toxic substance.

toxicity   An innate potential of a material to cause a harmful effect.

toxicity characteristic leaching procedure   [TCLP]   An EPA method to determine if a waste material meets the criteria for toxicity per the CERCLA requirements.

toxic material   A substance that may produce an injurious or lethal effect through ingestion, inhalation, or absorption through any body surface. *See* also toxic substance.

toxicology   The study of the harmful effects of chemicals on biologic systems. The quantitative study of the injurious effects caused by chemical and physical agents. The study of the nature and action of poisons. The study of the adverse health effects caused by chemicals in humans and animals.

Toxicology Information On-Line   [TOXLINE]   An on-line service (via telephone terminal) of the National Library of Medicine to provide current and prompt information on the toxicity of substances on request.

toxic substance (OSHA)   A substance that demonstrates a potential to produce cancer, to produce short-term and long-term disease or bodily injury, to affect health adversely, to produce acute discomfort, or to endanger the life of man or animal, as a result of exposure via the respiratory tract, skin, eye, mouth, or other route, in quantities that are reasonable for experimental animals or that have been reported to have produced toxic effects in man. A chemical or other substance that may present an unreasonable risk of injury to health or the environment.

Toxic Substances Control Act   [TSCA]   An act developed and implemented by the U.S. Environmental Protection Agency to regulate new chemicals entering the market.

toxic symptom   An indication of the presence of a toxic substance in the body.

toxin   A poisonous substance, having a protein structure, secreted by certain organisms and capable of causing a pathological response when introduced into the body.

TOXLINE   Toxicology Information On-line.

TP   Total pressure.

TPQ   Threshold Planning Quantity.

TPY   Tons per year.

TQM   Total quality management.

tracer (Gas)   A gas, such as sulfur hexafluoride, that can be used to identify suspected pollutant pathways and to quantify ventilation rates.

tracer (Radioactive)   An isotope, or non-natural mixture of isotopes of an element, that may be incorporated into a substance for determining metabolic pathways, mode of action, site of action, rates of excretion, etc.

trachea   The windpipe that conducts air to and from the lungs.

tracheitis   Inflammation of the trachea.

tracheotomy   A procedure in which an incision is made into the trachea through the neck.

track etch detector   A type of detector used to determine the presence and amount of radon gas contamination.

trade name   The commercial name or trademark name of a product or material.

trait   A distinguishing feature or characteristic.

transducer   A device, such as a photocell or piezoelectric crystal, that converts input energy of one form into output energy of another form.

transformation   The conversion of a chemical substance from one form to another.

transient   Something lasting only a short period of time.

transient sounds   Sounds whose average level does not remain constant with time.

transient vibration   The temporary vibration of a mechanical system.

transition (Ductwork)   A change in the cross-sectional shape or area of a duct or hood.

translucent   Quality of transmitting light but creating sufficient diffusion to prevent the perception of clear images of objects when looking through the translucent material.

transmission electron microscope   [TEM]   A microscope that utilizes an electron beam that is focused on a sample to produce an image showing differences in density of the sample material on a fluorescent screen from which the sample can be identified, and counted when fibrous.

transmission loss (Acoustics)   The reduction in noise when a diffuse noise field is generated on one side of a test panel (e.g., a side of a room) and the noise is determined on the opposite side of the panel. The transmission loss of the panel material and the measurement of the sound absorption of the room can be calculated from this data.

transmittance   The fraction of incident light that is transmitted through a medium of interest.

transplacental   The passage of materials, such as chemicals, across the placenta.

transplacental toxin   A substance that causes an adverse effect in a developing embryo.

transport index (Ionizing Radiation)   The dose rate in millirem per hour (mrem/hr) at 3 feet from the surface of a package containing radioactive materials.

transport velocity   The velocity needed to prevent the settling of airborne dusts or other particulates in the ductwork of a ventilation system.

transuranium elements   Nuclides having an atomic number greater than that of uranium (i.e., 92) and that are not found naturally, but are produced by nuclear bombardment. Also referred to as transuranic elements.

trauma   A wound or injury. Can also be due to stress from an external source.

TRC   Total residual chlorine.

tremor   An involuntary trembling motion of the body or a part of the body.

trench foot   Effect on the feet, resembling frostbite, as a result of prolonged standing, relatively inactively, with wet feet in a cold environment.

trend   Inclination in a particular direction or course.

triage   The procedure of sorting out and classifying casualties to determine the priority of need and proper place for treatment.

tridymite   A form of free silica formed when quartz is heated to 870°C.

trigger finger   Locking of a finger when in the extension or flexion (bent) position due to constriction of a tendon sheath.

troposphere   That portion of the atmosphere between 7 and 10 miles from the earth's surface.

TRQ   Threshold reportable quantity.

TSCA   Toxic Substances Control Act.

TSDF   Treatment, storage, and disposal facility.

TSI   Thermal system insulation.

TSP   Total suspended particulate.

TSS   Total suspended solids.

TTS   Temporary threshold shift.

tularemia   An infectious disease caused by a bacterium that is transmitted to man by insects or by the handling of infected animals. Also called rabbit fever.

tumor   A swelling or abnormal mass of tissue that may or may not be malignant. A new growth of tissue in which the multiplication of cells is uncontrolled and progressive. Also called a neoplasm.

tumorigenic agent   A substance that produces tumors.

tunnel vision   A narrow field of vision.

turbidimeter   A device used to measure the amount of suspended solids in a liquid.

turbidity   Unclear appearance of a fluid, such as air or water, due to the presence of suspended solids.

turbulence loss   Loss that occurs in a ventilation system whenever airflow changes direction or velocity and results in a pressure drop as the air flows through the ventilation system. Also referred to as dynamic loss.

turbulent flow   Fluid particles moving in directions other than a straight line parallel to the axis of the pipe or duct.

turbulent noise   Noise caused by air or gas moving through the fan and the transport (duct/pipe) system.

turnaround   A time period during which a plant is shut down for inspection, maintenance, repair, or modification that cannot be carried out while operating.

turning vane   Curved strips of short radii placed in a sharp bend in a duct or at a fan entry to direct air around the bend in a streamlined flow, thereby reducing turbulence losses.

turnover time (Indoor Air Quality)   Time equal to the effective room volume (i.e., volume of the space) divided by the ventilation airflow rate. Also known as air exchange time.

TVOC   Total volatile organic compounds.

TWA   Time-weighted average.

twitch   A brief contraction of a muscle.

tympanic cavity   The chamber of the middle ear.

tympanic membrane   *See* eardrum.

type C respirator   A respiratory protective device that is designed to provide protection to the wearer by providing clean air from a source outside the contaminated area.

U

ubiquitous   Being, or seeming to be, everywhere at the same time.

UEL   Upper explosive limit.

UFL   Upper flammable limit.

UHF   Ultra high frequency.

UL   Underwriters Laboratories, Inc.

ulcer   A local defect, or excavation, of the surface of an organ or tissue that is produced by the sloughing off of inflammatory necrotic tissue.

ulcerated   Afflicted with an ulcer.

ulceration   The formation of an open sore or ulcer.

ullage   The depth of the free space in a cargo tank above the liquid level.

ulnar deviation   A position of the hand in which the angle on the little finger side of the hand is decreased with respect to the corresponding side of the forearm. This working position may cause nerve and tendon damage.

ULPA filter   Ultra low penetration air filter.

ultrahigh frequency (Nonionizing Radiation)   Frequency in the range of 300 to 3000 hertz.

ultra low penetration air filter   A filter with a collection efficiency of 99.9995 percent.

ultrasonic   High frequency sound waves that are beyond the range of human hearing, which is generally considered to be 20,000 hertz.

ultrasonics   The acoustics of ultrasonic sound.

ultrasonic sound   Sounds in the frequency range above 20,000 hertz, that is, above the audible range.

ultraviolet radiation   Electromagnetic radiation with wavelengths from 10 to 380 nanometers.

unattended laboratory operation   A laboratory procedure or operation at which there is no person present who is knowledgeable regarding the operation and the emergency shut-down procedure.

underground injection   The placement of fluids underground through a bored, drilled, or driven well.

underground storage tank   [UST]   A storage tank located underground. Typically used to store gasoline (i.e., at gas stations), other oil products, or hazardous substances. There are requirements for depth and cover, corrosion protection, venting, etc.

Underwriters Laboratories, Inc.   [UL]   Independent, nonprofit organization that operates laboratories for examining and testing systems, devices, and materials of interest to the public safety.

undulant fever   Brucellosis. A persistent and recurring fever caused by a bacteria that is transmitted to man as a result of contact with infected animals or by consuming infected meat or milk.

uninterrupted power supply   [UPS]   A system consisting of a battery source, converter, inverter, and control equipment designed to provide a clean, conditioned, sinusoidal wave of electric power for a finite period of time.

union shop   A business or industrial plant whose employees are required to be union members or who agree to join the union within a specified time after being hired.

unit air conditioner   Factory-produced air conditioning unit serving one room or area.

United States Code   The official compilation of federal statutes.

United States Public Health Service   [USPHS]   *See* Public Health Service.

universal precautions (Bloodborne Pathogen)   An approach to infection control based on the concept of universal precautions, that all human blood and certain human body fluids are treated as if known to be infectious for HIV, HBV, and other bloodborne pathogens.

unleaded gasoline   Motor gasoline that does not contain a lead additive (e.g., tetraethyl or tetramethyl lead).

unrestricted area (Ionizing Radiation)   An area in which the radiation dose to a person would be less than 2 millirem in any 1 hour or 100 millirem per week.

unsafe act   Conduct that causes an unnecessary exposure to a hazard or a violation of a commonly accepted procedure that directly permits or results in a near miss or the occurrence of an accident.

unsafe condition   Any physical state that deviates from the accepted, normal, or correct practice, and that has the potential to produce injury, excessive exposure to a health hazard, or property damage.

unsaturated compound   An organic chemical compound that contains at least one double or triple carbon-to-carbon bond.

unstable (Ionizing radiation)   The quality of all radioactive materials in that they disintegrate spontaneously to form other elements.

unstable material   A material that, in the pure state or as commercially produced, will vigorously polymerize, decompose, condense, or become self-reactive and undergo other violent chemical changes.

upper explosive limit   [UEL]   The highest concentration of a vapor or gas in air that will produce a flash when an ignition source is present.

upper flammable limit   [UFL]   *See* upper explosive limit.

upper respiratory infection   An infection in the upper respiratory tract.

upper respiratory tract   The mouth, nose, sinuses, and throat above the epiglottis.

upper respiratory tract irritant   A substance that is irritating to the upper respiratory tract.

UPS   Uninterrupted power supply.

upset   The unanticipated malfunction of a process operation.

uptake (Toxicology)   The absorption of a substance into the body through the intestines, lungs, or skin.

uranium series   Isotopes which belong to a chain of successive decays that begins with uranium 238 and ends with lead 206.

uremia   The retention of excessive by-products of protein metabolism in the blood, and the toxic condition produced as a result.

URI   Upper respiratory infection.

urinalysis   The determination of the composition of urine, or the analysis for a specific substance in urine.

URT   Upper respiratory tract.

urticaria   Hives. A vascular reaction of the skin marked by the transient appearance of smooth, slightly raised patches that are attended by severe itching.

USC   United States Code.

USDA   United States Department of Agriculture.

useful beam (Ionizing Radiation)   Photons coming directly from the source, through collimating devices, and directly to the target to be irradiated.

user seal check (Respiratory Protection)   The procedure of checking to ensure that a good facepiece seal has been achieved. A positive or negative pressure check must be made every time a tight-fitting respirator is put on. *See* fit check.

USPHS   United States Public Health Service.

UST   Underground storage tank.

utensil   Any instrument or container, especially those in domestic use, used in the preparation and consumption of food.

U-tube   A see-through tube (e.g., glass or plastic) that is mounted on a scale for determining pressure using flexible tubing connected to a pitot tube or pressure tap on ductwork. Referred to as a U-tube manometer.

UV   Ultraviolet radiation/light.

UV–A   Long-wave ultraviolet radiation.

UV–B   Short-wave ultraviolet radiation.

UV detector   A detection system in which ultraviolet radiation is passed through a cell containing a sampled material. The absorption of ultraviolet energy at a wavelength that coincides with the absorption band of the analyte (contaminant) is proportional to the amount of contaminant in the sample. This can be used to calculate the concentration of the contaminant.

UVS   Ultraviolet spectrophotometry.

V   velocity.

V   Volt.

vaccinate   To provide active immunity by vaccination.

vaccination   The injection of a vaccine for the purpose of inducing immunity.

vaccine   A suspension of attenuated or killed microorganisms that is administered for the prevention, amelioration, or treatment of infectious diseases.

vacuum   A pressure less than that exerted by the atmosphere.

valence   A whole number representing or denoting the chemical combining power of one element with another; the number of electrons that can be lost, gained, or shared by an atom when combining with another element or ion.

valid   Supportable and verifiably correct.

validate   To verify or confirm.

validated method (Monitoring)   A sampling/analytical method that has been evaluated and determined to be effective for assessing worker's exposure to a contaminant. A method should have an efficiency of 75 percent at the 95 percent-confidence limit to be considered acceptable.

validation (Analytical Method)   A procedure to be followed to demonstrate that a method is sound and meets precision and accuracy parameters.

vapor   The gaseous form of substances that are normally in the solid or liquid state at normal temperature and pressure, and that can be changed to these states from the vapor state by increasing the pressure or decreasing the temperature.

vapor density   The weight of a given volume of pure vapor compared to the weight of an equal volume of dry air at the same temperature and pressure.

vapor/gas spiking   Sampling air through a media to which an analyte of interest has been added. Subsequent desorption and analysis provides information on sampling media effectiveness, analyte losses during sampling, sample stability, and other factors. A recovery of greater than or equal to 75 percent should be realized for the method to be considered acceptable.

vapor/hazard ratio   The ratio of the equilibrium vapor concentration at 25°C to the 8-hour time-weighted average TLV (i.e., ppm per ppm).

vaporization   The change of a substance from the liquid or solid state to the vapor state.

vapor phase   The existence of a substance in the gaseous state.

vapor pressure   [VP]   The pressure, usually expressed in millimeters of mercury, characteristic at any given temperature of a vapor in equilibrium with its liquid or solid form; the pressure exerted by a vapor. If a vapor is kept in confinement over its liquid so that the vapor can accumulate above the liquid, the temperature being held constant, the vapor pressure approaches a fixed limit referred to as the saturated vapor pressure, which is dependent only on the temperature and the liquid.

vapor recovery   A system or method by which vapors are retained and conserved.

vapor spike   A sorbent media sampling device to which a known amount of a substance, in the form of a vapor, has been added.

variable   A quantity that can vary.

variable air volume system   [VAV]   Air handling system that conditions air to a constant temperature and varies the outside airflow (make-up air) to ensure thermal comfort.

variance (Sample)   A measure of the spread or variability of the sample data; an estimate of the population variability equal to the average of the mean squares of the deviations from the arithmetic mean of a number of values of some variable.

variation   The act, process, or result of change.

varicose   Blood or lymph vessels that are abnormally dilated or knotted.

vascular   Pertaining to blood vessels.

vasoconstriction   The reduction in the opening of a vessel, especially constriction of arterioles that leads to a decrease in the blood supply to a part.

vasospasm   The spasm of blood vessels, resulting in a decrease in their opening.

VAT   Vinyl asbestos tile.

VAV   Variable air volume.

VDT   Video display terminal.

vector (Ergonomics)   A quantity that has both magnitude and direction.

vector (Public Health)   The vehicle by which an infectious agent is transferred from an infected to a susceptible host.

vector-borne disease   A disease that is transmitted from one host to another by a carrier, such as an arthropod.

vector control   The surveillance and control of rodents, insects, and other vectors of public health concern and economic significance.

vein   A vessel through which blood flows from various organs or parts of the body back to the heart.

velocity   A vector specifying the time rate of change of displacement with respect to a reference.

velocity contours (Ventilation)   *See* flow contours.

velocity of sound   The speed at which sound travels, 340 meters (1115 feet) per second.

velocity pressure   [VP]   The kinetic pressure in the direction of flow necessary to cause a fluid at rest to flow at a given velocity. Usually expressed in inches of water when the fluid is air.

velometer   A device for measuring air velocity.

vena contracta   The narrowing in the diameter of an air stream as it enters a duct, hood, or other part of a ventilation system.

venom   Complex chemical and pharmacologically diverse toxic secretions of animals that are transmitted to man by means of a bite, sting, puncture, or by spitting.

venomous   Venom-producing or secreting.

vent   An opening allowing the passage or escape of gas, steam, fumes, air, etc.

ventilate   To provide fresh air to an area by operating a fan or opening a window.

ventilation   A method for controlling health hazards by causing fresh air to replace contaminated air, which is simultaneously removed as the fresh air is introduced.

ventilation air (Indoor Air Quality)   The total air supplied to an area, which is the combination of air brought into the system from the outdoors and air that is being recirculated within a building.

ventilation effectiveness (Indoor Air Quality)   The fraction of outdoor air delivered to a space that reaches the occupied zone.

ventilatory volume   The volume of air exchanged per unit time between the lungs and the atmosphere.

venturi   A constriction in a section of pipe or duct to accelerate the fluid and lower its static pressure. A venturi can be used to determine fluid flow by determining the pressure difference between the pressure in the upstream pipe/duct and that in the constriction.

venturi meter   Typically a section of piping or duct with a contraction (25 degrees) to a throat with a re-expansion (7 degrees) to the original diameter/size. This device is used to measure mass flow rate based on an empirical formula.

venturi scrubber   An air cleaning unit that relies on the creation of an increase in air velocity of the exhaust created by a venturi, with the subsequent removal of particulates from the exhaust air stream in a cyclone separator.

vermiculite   A mineral with a platelet-like crystalline structure that is lightweight and highly water absorbent.

vermin   Various small animals or insects that are destructive, annoying, or injurious to health.

vermin control   A facility constructed, equipped, and maintained, so far as is reasonably practical, to prevent the entrance or harborage of rodents, insects, and other vermin.

verruca   Wart.

vertigo   Dizziness, or the sensation that the environment is revolving around a person.

vesicant   A substance that produces blisters on the skin.

vesicle   A small blister on the skin.

vesiculation   The presence of or the formation of vesicles.

VHF   Very high frequency.

viability   Ability to remain alive in a free state.

viable   Living.

viable count   In a sample, a count of the cells that are able to grow and reproduce.

vibration   A rapid linear motion of a particle or an elastic solid about an equilibrium position; the act of vibrating.

vibration isolator   A resilient support that tends to isolate a system from steady-state excitation.

vibration white fingers   *See* Raynaud's syndrome.

video display terminal   Computer screen-based terminal.

vinyl asbestos tile   [VAT]   A floor covering material that contains asbestos.

virgin material   A raw material.

virology   The study of viruses and viral diseases.

virulence   The degree of pathogenicity of a microorganism as indicated by case fatality rates and/or its ability to invade the tissue of the host.

virulent   Extremely poisonous or venomous; capable of overcoming bodily defensive mechanisms. Any infection marked by rapid onset of severe symptoms due to the toxic effects produced by the pathogen.

virus   A pathogen consisting mostly of nucleic acid and that lacks a cellular structure.

visceral   Relating to an internal organ, especially abdominal organs.

viscid   Thick and adhesive.

viscosity   A measure of a liquid's internal friction or of its resistance to flow.

viscous   Having high resistance to flow.

visibility   The distance, under existing weather conditions, that it is possible to see visually.

visible   Pertaining to radiant energy in the electromagnetic spectral range that is visible to the human eye.

visible emission   An emission, from a source, that is visually detectable without the aid of instruments.

visible light   Electromagnetic energy having wavelengths within the range of 380 to 770 nanometers.

visual acuity   The ability of the eye to sharply perceive the shape of objects in the direct line of vision.

vital capacity   [VC]   *See* forced vital capacity.

vitality   The property of being alive and vigorous.

vital signs   Measurements of signs of life, including temperature, pulse rate, respiratory rate, and blood pressure.

vital statistics   Data that record significant events and dates in human life, such as births, deaths, marriages, etc.

vitrification   The process in which high temperatures are employed to form glass from a ceramic and some mineral materials.

V/m   Volts per meter.

VM & P Naphtha   Varnish Maker's and Painter's Naphtha.

VOC   Volatile organic compound.

vol   Volume.

volatile   A substance that evaporates at a low temperature.

volatile organic compounds   [VOCs]   Organic compounds that can release vapors into the surrounding air. When the vapor is present in sufficient quantity, it can cause eye, nose, and throat irritation, headache, etc. Some volatile organic compounds, such as benzene, are of greater concern than others due to their known adverse health effects at higher concentrations.

volatility   The tendency or ability of a liquid to volatilize or evaporate.

volt   [V]   The unit of electromotive force.

volts per meter   [V/m]   A measure of electric field strength.

volumetric analysis   The measurement of the volume of a liquid reagent of known concentration that is required to react completely with a substance whose concentration is being determined. Titrations of acids with a base, or vice versa, are examples of this type of analysis.

voluntary risk   Risk that an individual has consented to accept.

vortex   Fluid flow involving the rotation of the fluid about an axis, such as the movement of air in a tornado.

vp   Vapor pressure.

VP   Velocity pressure.

VPP (OSHA)   Volunteer Protection Program.

VS   Visible spectrophotometry.

v/v   Volume to volume.

VWF   Vibration-induced white fingers. *See* Raynaud's syndrome.

W

W Watt(s).

walk-in hood A fume hood designed to be floor-mounted with sash and/or doors closing the open face.

warm-up time (Instrument) The period of time from when an instrument is turned on to the time when it will perform to its specifications.

warning A method of notification of a potential hazard and the recommended actions to take to reduce risk.

warning property That property of a substance that enables a worker to identify a potential excessive exposure situation while wearing a respirator or while in the work environment. If the odor threshold of a material is below the acceptable exposure limit, it can serve to alert a worker to the presence of a substance at an excessive level and to take appropriate measures to prevent an excessive exposure.

warning property (Respiratory Protection) A contaminant with an odor threshold below its permissible exposure limit does not provide adequate warning to wearers of air-purifying respirators if breakthrough occurs. Thus, air-purifying respiratory protection is not recommended for substances with poor warning properties. Some substances (e.g., hydrogen sulfide) can cause olfactory fatigue and, even though they have a low odor threshold (i.e., below their exposure limit), the use of air-purifying respirators is not recommended for protection against them.

wart A verruca.

waste Unwanted material from a process operation; refuse from an industrial or combustion operation as well as from animal or human habitation.

waste disposal The process or means for getting rid of waste material, such as in an approved landfill.

waste minimization The use of treatment processes that reduce the amount of waste requiring disposal.

waste reduction Cutting down the quantities of wastes generated at their sources.

wastewater Industrial or municipal wastewater.

water column A term used to express pressure (e.g., inches of water column or inches water gage).

water curtain A means for reducing or preventing the emission of paint during spray painting operations by providing a water flow over a wall located at the rear of a paint spray booth to collect paint overspray.

water gage *See* water column.

water hammer The effect of pressure rise that may accompany a sudden change in the velocity of water or other liquid flowing in a pipe. It often occurs if there is a sudden closure of a valve or stopping of a pump in a pipeline or in equipment handling the liquid.

water manometer A tube shaped in the form of a U that is used to measure pressure differences when properly attached to a duct, fan, hood, scrubber, etc.

water pollution   The release of a harmful or objectionable material, resulting in a degradation of the water quality.

water quality   The quality of water with respect to its intended usage.

water quality standards   Limits that have been established by various government agencies specifying concentrations of contamination that are acceptable, such as those for drinking water.

water reactive   Material that reacts with water, resulting in a safety or health hazard.

watershed   The area that drains into an outlet such as a river.

water solubility   The maximum amount of a substance that results when it is dissolved in water.

water table   The underground level at which water is found.

watertight (Instrument)   So constructed that moisture will not enter the instrument enclosure under specified test conditions.

watt   [W]   A unit of power in the International System equal to one joule per second.

wavelength   The distance, measured along the line of propogation, between any two points that are in phase on adjacent waves.

Wb   Weber.

WBGT   Wet-Bulb Globe Temperature.

weather cap   *See* rain cap.

weatherproof (Instrument)   So constructed or protected that exposure to the weather will not interfere with the instrument's operation.

weaver's cough   Acute respiratory illness that occurs among weaving mill employees as a result of their exposure to cotton dust.

weber   [Wb]   The International System unit of magnetic flux.

WEEL   Workplace environmental exposure level.

weight   The force with which a body is attracted toward the earth.

weighting (Acoustics)   The prescribed frequency response provided for in a sound-level meter (see A-, B- or C-weighted sound level).

weighting network (Acoustics)   Electrical networks (A,B,C) that are incorporated into sound-level meters. The C network provides a flat response over the frequency range of 20 to 10,000 hertz, while the B and A networks selectively discriminate against lower frequency sounds.

welder's flash   Eye effect (i.e., inflammation of the cornea) resulting from exposure to the UV radiation associated with arc welding. Also referred to as flashburn, or ground-glass eyeball.

welder's lung   A pneumoconiosis resulting from the deposition of particles in the lungs of welders.

welding fumes   Fumes generated during metal arc welding, oxy-acetylene welding, or other welding procedures in which iron, mild steel, or aluminum are joined; measured as total particulate in the breathing zone of the welder. OSHA recommends determining welding fume exposure by sampling inside the welding mask.

well injection   *See* underground injection.

wet-bulb depression   The difference between the temperature of the dry-bulb and the wet-bulb thermometers of a psychrometer.

wet-bulb globe temperature index   [WBGT]   A widely accepted index for determining heat stress on a worker in an indoor or outdoor environment.

wet-bulb temperature   The temperature at which liquid or solid water, by evaporating into the air, can bring the air to saturation at the same temperature; the temperature of air as measured by a wet-bulb thermometer, which is lower than the dry-bulb temperature except when air is saturated.

wet-bulb thermometer   A thermometer in which the bulb is covered with a wetted wick or cloth.

wet gas   A gas that contains water, or a gas that has not been dehydrated.

wet-gas meter   A meter for measuring gas volume. It is a secondary standard.

wet-globe temperature   The temperature determined using a wet-globe thermometer (e.g., Botsball).

wet heat   The use of steam to effect sterilization. It is a more effective method than using dry heat.

wet scrubber   A chamber through which exhaust air is passed and in which there are water sprays or bubble plates for removing particulates and soluble gases from exhaust air.

wet ice   Frozen water.

wet suit   A diving suit, usually made of neoprene material, designed to provide thermal insulation for a diver's body.

wetted black globe thermometer   [WGT]   such as the Botsball device.

wet-test meter   A secondary calibration device that can be used to determine the flow rate of a sampling pump.

wetting agent   A surface-active agent that, when added to water, causes it to penetrate more easily into or spread more easily over the surface of another material by reducing the surface tension of the water.

wg   Water gauge.

wgt   Wet-globe temperature.

wheal   A smooth, slightly elevated area on the body surface that is redder or paler than the surrounding skin.

Wheatstone bridge   An electrical circuit enabling the measurement of an unknown resistance by comparing it with a known resistance. It is the principle of operation of a number of combustible gas meters and other instruments.

wheeze   To breathe with difficulty, producing a hoarse whistling sound.

Whipple disc   A microscopic eyepiece with an inscribed grid that defines a specific area. Used in counting dust samples to determine the particle concentration.

white blood cell   A colorless cell of the blood also known as a leukocyte.

white damp   Carbon monoxide.

white finger disease   *See* Raynaud's syndrome.

white lung   Term used to indicate the effect of asbestos on the lungs.

white noise   A noise that is uniform in power-per-hertz-bandwidth over a very wide frequency range. *See* broadband noise.

WHMIS   Canadian regulation on Workplace Hazardous Materials Information System. Requires that information on hazardous materials in the workplace be provided to potentially exposed workers.

WHO   World Health Organization.

whole body dose (Ionizing Radiation)   The radiation dose received by the whole body and not just to a part.

whole body vibration   The vibration transmitted to the entire human body through a supporting structure, such as a vehicle seat or building floor.

willful violation (OSHA)   A violation of a standard in which the employer either knew that what was being done constituted a violation or was aware that a hazardous condition existed and made no reasonable effort to eliminate it.

willful violation citation (OSHA)   A citation issued if an employer committed an intentional and knowing violation of the OSHA Act, or issued when the employer was aware that a hazardous condition existed and did not make a reasonable effort to eliminate the condition.

Williams-Steiger Occupational Safety and Health Act   The act of Congress that created OSHA.

wind   The horizontal movement of air over the earth's surface.

windbox   A chamber below a furnace grate or burner, through which air is supplied for combusting the fuel.

wind noise   The low frequency response from wind impacting a microphone. The higher the wind velocity, the greater the effect. A wind screen is typically effective in eliminating the wind effect on a microphone.

wind rose   A polar diagram presenting information on wind direction and wind speed. Typically, wind direction is indicated by the length of spokes at 8 or 16 positions around a circle with north, east, south, and west identified, and with each spoke divided into segments of different lengths to indicate wind velocity.

wind screen   A covering for a microphone to reduce the effect of wind on microphone response.

wind shear   Variation of the horizontal wind speed and direction with height.

wind vane   A device for indicating the direction from which the wind is coming; that is, it points upwind.

windward   The direction from which the wind is coming; upwind.

wipe test   The collection of chemical, mineralogical, or radiological stressors from a surface onto a media, such as filter paper. A typical wipe area is 100 square centimeters. Wipe test results are useful indices of contamination, but not direct estimators of exposure risk.

wk   Week.

W/kg   Watts per kilogram.

WL   Working Level.

WLM   Working Level Month(s).

woolsorter's disease   Pulmonary anthrax.

work area   A specified area within a workplace where an individual or individuals perform assigned work.

worker's compensation   A type of insurance coverage that is required by law and compensates workers who have been injured on the job, regardless of fault.

work history   A historical representation of the various jobs and locations/areas where an individual worked over a working lifetime.

working alone   The performance of any work by an individual who is out of audio or visual range of another individual for more than a few minutes at a time.

Working Level   Any combination of short-lived radon decay products in 1 liter of air that will result in the ultimate emission of alpha particles with a total energy of $1.3\ E^5$ MeV.

Working Level Month   [WLM]   An exposure to 1 working level for 170 hours.

working standard   A solution prepared by volumetric dilution of a stock or intermediate solution and used directly to calibrate an instrument or to determine instrument response.

workmen's compensation   An insurance system that is required by state law and financed by employers. It provides payments to employees or their families for occupational illness, injuries, or fatalities that result in loss of wages or income while at work or loss of bodily function, regardless of whether the employee was negligent.

work permit   A document issued by an authorized person permitting specific work to be done during a specified period of time in a defined manner.

work physiology   A subdiscipline of ergonomics which addresses the effects of work on physiologic function, such as the assessmant of the capacity to perform physical work as well as the effects of fatigue on work performance.

workplace   An establishment or work site at a fixed location that has within its bounds one or more work areas.

workplace environmental exposure levels   [WEEL]   Exposure guides developed by the WEEL Committee of the AIHA for agents that have no current exposure guidelines established by other organizations. The WEELs represent time-weighted average workplace exposure levels to which, it is believed, nearly all employees could be repeatedly exposed without adverse effect.

workplace fit test   A respirator fit test procedure that is conducted at the workplace of the individual being provided the respiratory protective device.

workplace health hazard   Any material, physical agent, biological organism, or ergonomic stress for which there is evidence that acute or chronic health effects may result from exposure to it.

workplace protection factor   [WPF]   A measure of the actual protection provided in a workplace under the conditions of the workplace by a properly functioning respirator when correctly worn and used; the ratio of the measured time-weighted average concentration of the contaminant taken simultaneously inside and outside the respirator facepiece.

work practice controls   Methods used to prevent the release of a substance or physical agent in order to reduce the likelihood of exposure to it or contact with it. They prohibit certain actions by identifying specific ways to carry out a task and follow good personal hygiene practices. These can be applied to situations in which there is potential exposure via inhalation, skin contact, or skin absorption, as well as for exposures to physical agents (e.g., ionizing radiation, noise, heat stress, etc.).

worksite   A single physical location where business is conducted or operations are performed by the employees.

work tolerance (Ergonomics)   A condition in which a worker performs at an acceptable rate ergonomically, while experiencing both physiologic and emotional well-being.

World Health Organization   [WHO]   An international organization of a large number of countries that is operated within the United Nations Charter with the objective of working together to achieve a high level of health for the peoples of the world.

WPF   Workplace protection factor.

wt   Weight.

w/v   Weight per volume.

WWT   Wastewater treatment.

x-axis   The horizontal axis of a two-dimensional Cartesian coordinate system.

XDIF   X-ray diffraction.

xenobiotic   A chemical that is foreign to a biologic system.

xeroderma   Dry skin that may be rough as well as dry.

XF   X-ray fluorescence.

X-ray diffraction   [XRD]   An analytical method for identifying and quantifying crystalline materials by determining the diffraction pattern (diffraction beams and intensity) emitted by a material when exposed to X-rays.

X-ray fluorescence   [XRF]   An analytical method for identifying and quantifying the elements present in solids and liquids by examining the X-rays emitted (pattern and intensity) as a result of the absorption of radiation from some source (X-ray, isotope). This methodology is rarely used for analysis of gases.

X-rays   Highly penetrating ionizing radiations similar to gamma rays that are produced by bombarding a metal target with fast electrons in a vacuum tube, as occurs in an X-ray machine. X-rays are electromagnetic radiation at wavelengths shorter than those of visible light.

X-ray tube   An electron tube that is designed for the conversion of electrical energy into X-ray energy.

XRD   X-ray diffraction.

XRF   X-ray fluorescence.

Y

y.   Year.

yard   The fundamental unit of length in the U.S. and British system equal to 0.9144 meter.

y-axis   The vertical axis of a two-dimensional Cartesian coordinate system.

yd   Yard.

yd²   Square yard(s).

yd³   Cubic yard(s).

yellow fever   An acute infectious disease caused by a filterable virus transmitted by a mosquito.

youngest air (Indoor Air Quality)   That air found where the supply containing outdoor air enters a room.

yr.   Year.

YTD   Year to date.

Z

z   Symbol for atomic number.

zero air   Air containing no components other than those present in pure air.

zero discharge   A condition in which no waste that results in a discharge is produced.

zero drift (Instrument)   The change in instrument output over a stated period of unadjusted, continuous operation when the input concentration is zero. Drift in the zero indication of an instrument without any change in the input variable (i.e., contaminant).

zero gas (Instrumentation)   A gas that contains no contaminant that will cause a response in an instrument; enables the instrument response to be set at zero.

zero population growth   Limiting population growth or increase to the number of live births needed to replace the existing population.

zero threshold toxicant   *See* threshold toxicant.

Z-List   OSHA's tables (Z-1, Z-2, and Z-3) of Toxic and Hazardous Substances listing air contaminants and their applicable Permissible Exposure Limits in 29 CFR 1900.1000.

zinc metal fume fever   *See* metal fume fever.

zinc protoporphyrin   [ZPP]   *See* porphyrin.

zoonosis   A disease of animals that can be transmitted to man.

zootoxin   A toxic substance of animal origin, such as the venom of snakes, spiders, scorpions, etc.

REFERENCES

Ahlborn, A., *Biostatistics for Epidemiologists,* Lewis Publishers, Chelsea, MI, 1993.

American Chemical Society, *Chemical Hazards in the Workplace,* Washington, D.C., 1981.

American Conference of Governmental Industrial Hygienists, *A Guide for Control of Laser Hazards,* 4th ed., Cincinnati, OH, 1990.

American Conference of Governmental Industrial Hygienists, *Air Sampling Instruments for Evaluation of Atmospheric Contaminants,* 8th ed., Cincinnati, OH, 1995.

American Conference of Governmental Industrial Hygienists, *Guide to Occupational Exposure Values—1998,* Cincinnati, OH, 1998.

American Conference of Governmental Industrial Hygienists, *Industrial Ventilation,* 23rd ed., Cincinnati, OH, 1998.

American Conference of Governmental Industrial Hygienists, *Threshold Limit Values and Biological Exposure Indices,* Cincinnati, OH, 1998.

American Industrial Hygiene Association, *Biohazards Reference Manual,* Fairfax, VA, 1985.

American Industrial Hygiene Association, *Heating and Cooling for Man in Industry,* 2nd ed., Akron, OH, 1975.

American Industrial Hygiene Association, *Noise and Hearing Conservation Manual,* 4th ed., Fairfax, VA, 1986.

American Industrial Hygiene Association, *Quality Assurance Manual for Industrial Hygiene Chemistry,* Fairfax, VA, 1988.

American National Standards Institute, *Safety Requirements for Confined Spaces,* ANSI Z117.1–1989, New York, 1989.

American Petroleum Institute, *Guideline for Work in Confined Spaces in the Petroleum Industry,* Publ. Nos. 2217 and 2217A, Washington, D.C., 1984, 1987.

American Society of Heating, Refrigeration, and Air Conditioning Engineers, *Ventilation for Acceptable Indoor Air Quality,* ASHRAE 62–1989, Atlanta, 1989.

American Society for Testing and Materials, *Quality Assurance for Environmental Measurements,* Philadelphia, 1985.

American Welding Society, *Welding Safety and Health,* Miami, 1983.

Armour, M.A., *Hazardous Laboratory Chemicals Disposal Guide,* CRC Press, Boca Raton, FL, 1991.

Bearg, D.W., *Indoor Air Quality and HVAC Systems,* CRC Press, Boca Raton, FL, 1993.

Benenson, A.S., Ed., *Control of Communicable Diseases Manual,* American Public Health Assn., Washington, D.C., 1995.

Berger, E.H., Ward, W.D., Morrill, J.C., and Royster, J.H., Eds., *Noise and Hearing Conservation Manual,* 4th ed., American Industrial Hygiene Association, Fairfax, VA, 1988.

Bowman, V.A., Jr., *Checklists for Environmental Compliance,* Pudvan Publishing, Northbrook, IL, 1988.

Burgess, W.A., Ellenbecker, M.J., and Treitman, R.T., *Ventilation for Control of the Work Environment,* John Wiley & Sons, New York, 1989.

Burton, D.J., *IAQ and HVAC Workbook,* IVE, Bountiful, UT, 1993.

Burton, D.J., *Industrial Ventilation—A Self Study Companion to the ACGIH Ventilation Manual,* ECE, Salt Lake City, UT, 1982.

Chelton, C., Ed., *Manual of Recommended Practice for Combustible Gas Indicators and Portable Direct-Reading Hydrocarbon Detectors,* 2nd ed., American Industrial Hygiene Association, Fairfax, VA, 1993.

Cheremisinoff, P., *Hazardous Material—Manager's Desk Book On Regulation,* Pudvan Publishing, Northbrook, IL, 1986.

Clayman, C.B., Ed., *The American Medical Association Encyclopedia of Medicine,* Random House, New York, 1989.

DiBerardinis, L., et al., *Guidelines for Laboratory Design,* 2nd ed., John Wiley & Sons, New York, 1992.

Dorland's Illustrated Medical Dictionary, 28th ed., W.B. Saunders, Philadelphia, 1994.

Eisenbud, M., *An Environmental Odyssey,* University of Washington Press, Seattle, 1990.

Fay, B.A. and Billings, C.E., *Index of Signs and Symptoms of Industrial Diseases,* National Institute for Occupational Safety and Health, Cincinnati, OH, 1980.

Finucane, E.W., *Definitions, Conversions, and Calculations for Occupational Safety and Health Professionals,* Lewis Publishers, Boca Raton, FL, 1993.

Frick, W., Ed., *Environmental Glossary,* 3rd ed., Government Institutes, Rockville, MD, 1984.

Furr, A., *Handbook of Laboratory Safety,* 3rd ed., CRC Press, Boca Raton, FL, 1989.

Gold, D.T., *Fire Brigade Training Manual,* National Fire Protection Association, Quincy, MA, 1982.

Hallenbeck, W.H., *Quantitative Risk Assessment for Environmental and Occupational Health,* Lewis Publications, Chelsea, MI, 1993.

Handbook of Compressed Gases, 2nd ed., Van Nostrand Reinhold, New York, 1981.

Health Physics and Radiological Health Handbook, Nucleon Lectern Associates, Olney, MD, 1984.

Higgins, T.E., *Hazardous Waste Minimization Handbook,* Lewis Publishers, Chelsea, MI, 1989.

Illuminating Engineering Society, *IES Lighting Handbook,* 5th ed., New York, 1972.

Johnson, J.S. and Anderson, K.J., Eds., *Chemical Protective Clothing,* Vols. 1 and 2, American Industrial Hygiene Association, Fairfax, VA, 1990.

Koren, H., *Environmental Health and Occupational Safety,* CRC Press, Boca Raton, FL, 1996.

Leecraft, J., *A Dictionary of Petroleum Terms,* 3rd ed., Petroleum Extension Service, University of Texas, Austin, TX, 1983.

Levy, B.S. and Wegman, D.H., Eds., *Occupational Health—Recognizing and Preventing Work Related Disease,* 2nd ed., Little, Brown, Boston, 1988.

Lipton, S. and Lynch, J.R., *Health Hazard Control in the Chemical Process Industry,* John Wiley & Sons, New York, 1987.

Manahan, S.E., *Environmental Chemistry,* 6th ed., CRC Press, Boca Raton, FL, 1994.

Matheson Gas Data Handbook, 6th ed., Matheson, East Rutherford, NJ, 1980.

McNair, H.M. and Bonelli, E.J., *Basic Gas Chromatography,* Varian Aerograph, Walnut Creek, CA, 1987.

Molak, V., *Fundamentals of Risk Analysis and Risk Management,* CRC Press, Boca Raton, FL, 1997.

National Council on Radiation Protection and Measurements, *SI Units in Radiation Protection and Measurements,* NCRP Report No. 82, Bethesda, MD, 1985.

National Fire Protection Association, *Fire Protection Handbook,* 17th ed., Quincy, MA, 1986.

National Fire Protection Association, *National Fire Codes,* Vols. 1, 2, 3, 5, 6, 7, 11, Quincy, MA, 1991.

National Institute for Occupational Safety and Health, *Guide to Industrial Respiratory Protection,* Cincinnati, OH, 1987.

National Institute for Occupational Safety and Health, *Introduction to Occupational Health,* Cincinnati, OH, 1979.

National Institute for Occupational Safety and Health, *NIOSH Manual of Analytical Methods,* 4th ed., Cincinnati, OH, 1997.

National Institute for Occupational Safety and Health, *Pocket Guide to Chemical Hazards,* Cincinnati, OH, June 1997.

National Safety Council, *Accident Prevention Manual for Industrial Operations,* 8th ed., Chicago, 1980.

Olishifski, P.E., Ed., *Handbook of Hazardous Materials,* 2nd ed., Alliance of American Insurers, Schaumburg, IL, 1983.

Patty's Industrial Hygiene and Toxicology, Vols. 1A and 1B, 4th ed., John Wiley & Sons, New York, 1991.

Peterson, R.D., and Cohen, J.M., *Complete Guide to OSHA Compliance,* CRC Press, Boca Raton, FL, 1996.

Phlog, B.A., Ed., *Fundamentals of Industrial Hygiene,* 3rd ed., National Safety Council, Chicago, 1991.

Prudent Practices for Handling Chemicals in Laboratories, National Academy Press, Washington, D.C., 1981.

Scott, R.M., *Introduction to Industrial Hygiene,* CRC Press, Boca Raton, FL, 1995.

Shapiro, J., *Radiation Protection—A Guide for Scientists and Physicians,* 3rd ed., Harvard University Press, Cambridge, MA, 1990.

Triola, M.F., *Elementary Statistics,* Benjamin/Cummings, Menlo Park, CA, 1983.

Turk, A., Turk, J., Wittes, J.T., and Wittes, R.E., *Environmental Science,* W.B. Saunders, Philadelphia, 1978.

U.S. Department of Health, Education, and Welfare; National Institute for Occupational Safety and Health, *Occupational Exposure Sampling Strategy Manual,* Cincinnati, OH, 1977.

U.S. Department of Health, Education, and Welfare, *Radiological Health Handbook,* Rockville, MD, 1970.

U.S. Environmental Protection Agency/National Institute for Occupational Safety and Health, *Building Air Quality,* Washington, D.C., 1991.

U.S. Environmental Protection Agency, *Managing Asbestos in Place,* Washington, D.C., 1990.

U.S. Government Printing Office, *Code of Federal Regulations—Title 29,* Parts 1900, 1910, and 1926, Washington, D.C., 1998.

Video Displays, Work, and Vision, National Academy Press, Washington, D.C., 1983.

Webster's Ninth Collegiate Dictionary, Merriam Webster, Springfield, MA, 1989.

University of Massachusetts Lowell Libraries